The GREAT UNSOLVED MYSTERIES of SCIENCE

The GREAT UNSOLVED MYSTERIES of SCIENCE

John Grant

CHARTWELL BOOKS, INC.

A QUINTET BOOK

Published by Chartwell Books
A Division of Book Sales, Inc.
110 Enterprise Avenue
Secaucus, New Jersey 07094

ISBN 1-55521-562-9

This book was designed and produced by
Quintet Publishing Limited
6 Blundell Street
London N7 9BH

Creative Director: Peter Bridgewater
Art Director: Ian Hunt
Designer: Annie Moss
Project Editor: Caroline Beattie
Editor: Susan Baker
Picture Researcher: Liz Eddison
Illustrator: Lorraine Harrison

Typeset in Great Britain by
Central Southern Typesetters, Eastbourne
Manufactured in Hong Kong by
Regent Publishing Services Limited
Printed in Hong Kong by
Leefung-Asco Printers Limited

CONTENTS

X
Z
D
H
O
C
H—C—OH
$CH_2OPO_3^{2}$

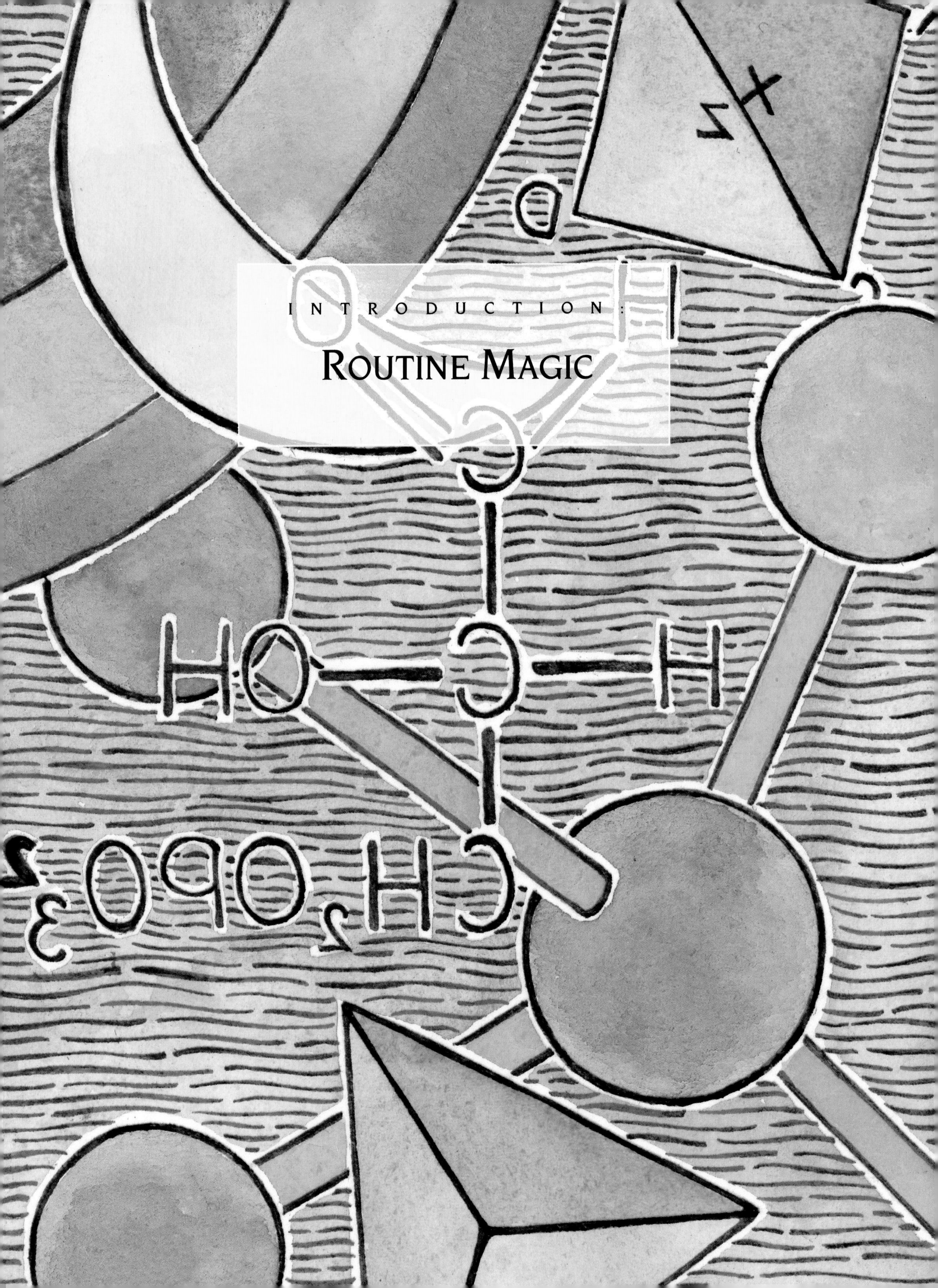

INTRODUCTION:

Routine Magic

Arthur C Clarke once remarked that any sufficiently advanced technology is, to the rest of us, indistinguishable from magic. To see the truth of this we need only look around us. For example, I am producing these words on a word-processor which shares my life for several hours each day. I know how to make it do all kinds of things – including a few that are not in the manufacturer's manual – but I have not the first idea as to how it actually works. To me, the whole process might just as well be magic.

Science is a form of magic, too – indeed, scientists at NASA, when performing yet another theoretically impossible feat, commonly refer to it as 'routine magic'. So I make no apologies for beginning this book in the way that a conjurer might, by giving you an apparently free choice from the pack while in fact forcing on you the particular card that I want you to take.

Pick a mathematical equation – any equation. What a surprise! You happen to have chosen this one:

$$x^n + y^n = z^n$$

Now this is a very interesting equation indeed. If *x*, *y*, *z* and *n* all represent whole numbers, it appears that the equation is an impossible one for any value of *n* greater than 2. The word 'appears' is used advisedly because, although computers have been used to show that the equation does not work for values of *n* up to several thousand, no one has yet been able to prove for certain that there is *no* number *n* greater than 2 for which suitable values of *x*, *y* and *z* cannot be slotted in to produce a valid equation.

This is something of a mystery – but there is a greater mystery involved. The Frenchman Pierre de Fermat (1601–65) was one of the greatest mathematicians ever to have lived: his contributions are too many to mention. Unfortunately, he was interested in mathematics only as a hobby, and so he did not bother to publish any of his work. (It first became known to the wider world when, five years after Fermat's death, his son published his notes.)

LEFT *Home computers were unheard-of only a couple of decades ago; now they have become 'routine magic'.*

One day Fermat was reading a book on mathematics when he had an inspiration. Hastily he scribbled in the margin,

> WHERE n IS A NUMBER LARGER THAN 2, THERE ARE NO WHOLE NUMBERS x, y, z SUCH THAT $x^n + y^n = z^n$, AND OF THIS I HAVE FOUND A MOST MARVELLOUS PROOF, BUT THIS MARGIN IS TOO SMALL TO CONTAIN IT.

And that is precisely *all* we have of what has now come to be known as Fermat's Last Theorem. We can discount the possibility that Fermat was lying: this was a personal note, for him alone to read. It is of course possible that he was wrong in his initial assumption – which would explain why he never expanded his 'proof' in his notes – but we have to remember that the man was a mathematical genius. For more than three centuries mathematicians have struggled to rediscover Fermat's Last Theorem, but without success. The matter is still a mystery.

Many of the mysteries of science are more important than Fermat's Last Theorem. Or are they? Even the most seemingly trivial gap in our knowledge or understanding can represent some very major failure in our comprehension – it can mean that our overall ideas are seriously wrong. This was shown dramatically in the early years of this century.

The planet Mercury orbits the Sun more closely than do any of the others. It is a small, rocky body; its surface is covered with craters and blistered by the heat of the Sun. Like the other planets it travels around the Sun in an ellipse, which means that at some times it is closer to the Sun than at others. The point of closest approach is called the *perihelion* of a planet.

All of this seemed to be well understood until early this century, because it was explained in terms of Newton's theory of gravitation. There was only one fly in the ointment – but it was the smallest, most insignificant of flies.

Mercury's perihelion *advanced* a little more than the theory said it ought to. This was discovered in about 1840 by the French astronomer Urbain Leverrier (1811–77).

When we draw a picture of a planet's orbit we show the Sun, of course, and a single line around it in the shape of an ellipse. (Imagine a 'squashed circle' with the Sun just off-centre.) However, the truth is not quite as simple as this, because the planet does not exactly retrace its path each time it goes around the Sun. Instead, the orbit as a whole twists a little further round each time, so that a true drawing of the planet's course should really look more like one of the patterns produced by a children's Spirograph toy. The net effect is that, each time, the perihelion is a little further round than the last time – or, to put it another way, the perihelion advances. We *now* know this to be true of all the planets; however, the effect is so small that only in the case of Mercury was it detectable by nineteenth-century astronomers. And even with Mercury the change involved is minuscule: it is, in terms of angles, about 40 seconds per century (there are 3600 seconds in each degree of arc).

ABOVE *Pierre de Fermat, the mathematical genius who left us with one of science's greatest conundra.*

For several decades most scientists assumed that the discrepancy was too negligible to worry about: after all, no one said that everything in the Universe should work perfectly. Others did worry, though, and agreed with Leverrier that there must be a planet even closer to the Sun than Mercury; the gravitational tug of this hypothetical planet – which Leverrier christened Vulcan – could cause the anomalous advance of Mercury's perihelion. Astronomers made strenuous efforts to observe Vulcan, generally attributing their lack of success to the fact that a small body so close to the Sun would be incredibly difficult to detect. The whole affair seemed merely a minor mystery.

In fact it was a major mystery. Realization dawned only in 1915, when a comparatively young theoretical physicist called Albert Einstein published the paper that is now generally called the General Theory of Relativity. This theory, in passing, exactly accounted for the advance of Mercury's perihelion. More importantly, it rewrote large chunks of accepted science. Without Einstein's insight, in part inspired by the 'trivial' matter of Mercury's orbit, our knowledge of science, not to mention our technology, would be at a much lower level than it is today.

Apart from anything else, it is unlikely that anyone would have been able to develop the various forms of 'magic' that allow my word-processor to work.

This book collects together what I consider to be some of the major unsolved mysteries of science. Of course, in making my selection I am almost certainly falling into the same trap as those nineteenth-century scientists who thought that the anomalous orbit of Mercury was interesting but, in essence, a matter of little concern. At the same time, I can guarantee that some of the unsolved questions are very important indeed. There could hardly be more fundamental mysteries than the reasons why the Universe came into existence, how life began on planet Earth, and so on. Other conundra may appear rather less 'cosmically' relevant; they may seem to be of little importance yet, like Mercury's orbit, may prove in the long term to be extremely important.

Perhaps arbitrarily, this book is divided into three parts dealing, respectively, with mysteries from the past, mysteries of life, and mysteries of physics. The first section

ABOVE *Stan Gooch, the author of* Cities of Dreams.
RIGHT *Religion can be regarded as a primitive attempt to explain scientifically all the phenomena around us. This Sri Lankan Buddhist priest confronts an image of the Buddha hoping to find, through contemplation, answers to the many mysteries humanity may never find the answer to.*

essentially deals with the sciences of geology and archaeology; the second with biology, sociology, psychology and anthropology; and the third with astronomy and physics. Of course, there is no real hard-and-fast barrier between these sciences: an advance in our understanding of physics can have profound effects on our understanding of biology, and so on. This interrelation of various disciplines is fundamental to science: the word 'science' itself derives from a Latin word meaning 'knowledge' – a term that embraces far more than the circumscribed disciplines just mentioned. It seems to be part of human nature, however, to seek to categorize things, and the various aspects of what is in essence a single, broad-based quest for further knowledge and understanding have not been spared.

Here, too, we have a mystery; this time a psychological one, or perhaps it is really the province of anthropology, or archaeology, or ... Why is it that we human beings are so *curious* about everything? Why is it that we should want to know things that are, in the broad scheme, totally irrelevant? You yourself are an example of this phenomenon at work, through the simple fact that you are reading this book. Could you not equally have said: 'Ha! Mysteries of science! Leave 'em to the scientists!'

Yet even people with severe mental retardation show a high level of curiosity in certain circumstances. As do higher-order animals, as anyone who lives with a cat or a dog will confirm. As do – but surely I cannot mean it! – lugworms. Lugworms are among the least intellectually blessed of all God's creatures, yet experiments have shown that they can (at least apparently) become bored, and the corollary of boredom is surely curiosity. This would suggest that curiosity is not a by-product of intelligence, as we might expect, but rather an essential property of life. Alternatively, it may be a prerequisite for the evolutionary development of intelligence – which leads us to the premonition that, in a few billion years' time, there may be intelligent lugworms stalking the Earth. The idea hardly bears thinking about.

Or perhaps it does. Chimpanzees are very intelligent – for animals. They have a highly developed sense of curiosity. Over the past few decades researchers have shown that chimps can create artworks (abstract thought), use tools in a quite sophisticated fashion, and understand the concept of language: humans and chimps can use sign languages like Ameslan to communicate with each other to a reasonably advanced level. (Interestingly, the chimps are able to manipulate the language to such an extent that they can create new words and phrases.) For all of these reasons, it hardly seems to offend rationality to assume that, over millennia or billennia, chimps will evolve to become as intelligent as we are now.

The lugworms may be lagging a long way behind, but perhaps their time may come?

Curiosity is a prime example of a mystery of science. It has particular interest in that we all know what the word means and yet none of us know what curiosity actually *is*. We recognize it when we come across it. The same is true, for that matter, of the phenomenon of intelligence – for which no one has as yet been able to produce an adequate definition.

We can assume – indeed, we *have* to assume – that our primitive forebears had a strong sense of curiosity. It is likely that some at least of nature's experiments as it sought to produce *Homo sapiens* were as intelligent as us, even though their type of intelligence may have been very different. (As an aside, Stan Gooch, in his 1989 book *Cities of Dreams*, presents the idea that Neanderthal Man – or, in his context, Neanderthal Woman – created a civilization as complex as ours but based on thoughts rather than on artefacts. His theory is controversial, to put it mildly; but at the very least he portrays the kind of different, 'alien' intelligence our ancestors may have had.) Those people of prehistory differed from us in one important respect: they did not possess the huge databank of past discoveries which we do. For example, if no one has yet discovered the principle of the lens, it is difficult to invent the telescope; if you want to erect an astronomical observatory, therefore, you have to do your best with the current technology – in other words, build Stonehenge or some similar megalithic monument. Your reasons for this colossal exploitation of what is to you hi-tech may be quite divorced from ours (*we* want to look for black holes, whereas *you* want to ascertain the precise time you should sacrifice a virgin in order to propitiate the gods), but the principle is much the same. You have, like it or not, curiosity and with it an adjunct: the desire to *explain* things. But perhaps these explanations may be a little bit too versatile for modern tastes:

- Why was there an eclipse? – The gods did it.
- Why did our village catch fire? – The gods did it.
- Why have we been flooded? – The gods did it.
- Why do I exist? – The gods did it.
- What will happen to me after I die? – It is in the lap of the gods.

Nevertheless, the very fact that an explanation should have been created at all tells us quite a lot about the way that the human mind works. The urge to explain seems to be part of the whole process that involves also the phenomena of curiosity and intelligence. This is why all those scientists have given so much time and effort to such apparent trivia as Fermat's Last Theorem and the advance of Mercury's perihelion. It seems that as a species, we can never be content until we have an explanation for *everything*.

Clearly there are many benefits to this urge. However, there is a negative side. Once an explanation – any explanation – has been produced, people tend to stick with it, no matter how much evidence thereafter appears to contradict it. This tends to happen at several levels. Our hypothetical cavedwellers with their universal explanation that 'the gods did it' are not too far separated from the modern Westerner who believes that the Soviet Union is an unalleviatedly 'evil empire' or that Blacks are inferior to Whites: all three explanations are easy to take on board, they are much simpler than the arguments put forward by the people who disagree with them, and, once accepted, they are clung to with a limpet-like grip.

Of course, people who have such simplistic ideas are never 'us'; they are always 'someone else'. They are the unfortunate primitives or fools: *we* know much better than that!

Do we? Most readers of this book aged over 30 will have been taught at school that the atom is made up of little billiard balls: in the centre there is the nucleus, made of biggish billiard balls called protons and neutrons; around the nucleus travel very small billiard balls called electrons. In fact, the atom is nothing like this (for example, electrons

BELOW *In a small Zimbabwean church built of straw a minister of God attempts to unravel the mysteries of the cosmos for the benefit of his congregation of women and children. Humanity has an intense desire to explain reality, and uses religion as a means of doing so.*

RIGHT *A Tunisian Rabbi seeks the answer to universal questions in a copy of the Bible.*

RIGHT *It's all just 'routine magic' to the little girl whose words spoken into the microphone call up a coloured image on screen. Adults, however, have greater difficulty in accommodating their views to the 'miracles' of modern technology.*

what is widely regarded as fact. Some of the ideas which I discuss are currently rated 'improbable', but should not be dismissed for that reason alone: their supporters may be in a minority, but that minority is often a very distinguished one.

A final note. People often treat science as if it were some sort of 'forbidden country'. In fact, one of the perennial irritations of those involved in the sciences is that it is somehow chic to be innumerate and totally ignorant of physics, mathematics and the rest, but at the same time taboo to be ignorant of the basics of, say, literature. This is not helped by an educational system and a cultural climate which often present science in a negative way – as either boring or difficult (or both).

In fact, the essential concepts of science are very often neither. I am aware, though, that when some readers first open this book they may find a few of the ideas discussed superficially intimidating. I would advise such readers not to lose heart: there is nothing in this book which cannot be grasped by an 11-year-old who employs a little application: I know, because I've tested it on an 11-year-old.

This book would be many times its current length had I explained everything in detail. In other words, in order to get across a general point I have often simplified the argument. I must ask those who are more familiar with the sciences to forgive me for any passages where they feel I might be guilty of *over*simplification. I would guide such readers to the 'Suggested Further Reading' section at the end of the book.

The use of the word 'unsolved' about any aspect of science implies that, somewhere, a solution awaits us. That is the glorious challenge of science.

PART ONE

Mysteries From The Past

s LP Hartley remarked in his classic novel *The Go-Between* (1953), 'the past is a foreign country: they do things differently there' Hartley was not, of course, referring to prehistory – except perhaps the prehistory of the individual soul – but his remark sums up our dilemma. When scientists attempt to unravel the mysteries of the past they always run up against a brick wall. They can take artefacts, fragments of bone, curious edifices and so on, and make inferences from these, but they know that they can never, ever state precisely the truth of the past. Perhaps if someone invented a time machine (almost certainly a scientific impossibility) archaeologists would have the satisfaction of being able to prove and disprove each other's theories; failing that, we have to accept that the science of the distant past is a matter of informed deduction and, let us face it, guesswork.

That this is the case is exemplified by several works of fantasy and science fiction. In more than one story Clifford Simak put forward the notion that the flesh-tones of dinosaurs might have been iridescent, rather than the drab grey-greens and browns depicted in so many artists' impressions. Of course, Simak was not putting this forward as a serious theory. He was simply pointing out that palaeontology, as a science, depends on the interpretation of fossil relics of bones, not flesh; no palaeontologist has any clue as to what a living dinosaur actually looked like. A similar point was made by the eminent US naturalist Stephen Jay Gould in his introduction to Bjørn Kurtén's fascinating novel *Dance of the Tiger* (1980), whose focus is the hypothetical extermination of

LEFT *Our ideas of the appearance of our prehistoric ancestors change according to current social attitudes. Life may indeed have been nasty, brutal and short a few hundred thousand years ago, but there are those who claim that at least the shortness was a benefit in comparison with modern human existence. The truth is that we do not even know what our ancestors actually looked like.*

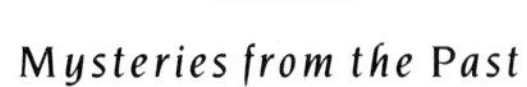

Neanderthals by Cro-Magnons, about 35,000 years ago. Gould remarks:

> ... I FELT ENLIGHTENED – AND EMBARRASSED. KURTÉN DEPICTS NEANDERTHALS AS WHITE-SKINNED, CRO-MAGNONS AS DARK. UNTIL READING THIS, I HAD NEVER REALIZED THAT MY UNQUESTIONED PICTURE OF NEANDERTHALS AS DARK AROSE FROM STANDARD RECONSTRUCTIONS ... AND FROM THE RACISM THAT SADLY AFFLICTS US ALL AND LEADS WHITE PEOPLE TO ASSOCIATE INFERIORITY AND DARKNESS. KURTÉN'S RECONSTRUCTION MAKES MORE SENSE, SINCE NEANDERTHALS LIVED IN GLACIAL ENVIRONMENTS AND LIGHT SKIN MAY BE AN ADAPTATION TO LIFE IN MIDDLE TO HIGH LATITUDES. YET ANY SCIENTIST WOULD BE RIGHTLY DUBBED A FOOL IF HE PUBLISHED A PROFESSIONAL PAPER ON 'THE SKIN COLOUR OF NEANDERTHAL DEDUCED FROM GENERAL EVOLUTIONARY PRINCIPLES'.

The complement of Gould's last comment is likewise true: any scientist who claimed that the Neanderthals were dark-skinned would clearly be committing an act of intellectual folly. The stark fact is that, as with the dinosaurs, we simply do not know what Neanderthal and Cro-Magnon people actually *looked* like. We know their average heights, but we do not know if our ancestors were hairy or smooth, blue-eyed or brown-eyed, loquacious or capable of communicating only the most fundamental of information in grunts. We do not know if their society (and here, of course, we are assuming that they actually *had* a society) was matriarchal or patriarchal. We do not know if they had music or even if they had genuine art: cave-paintings and bone-carvings might be manifestations of the muse, but they could equally well have been executed solely for mystical purposes.

In this section we shall look at some scientific mysteries that are rather more profound than the skin-colours of dinosaurs or prehistoric human beings. Instead, let us start with a fairly fundamental question ...

How Did Life Begin?

You may not seem to have much in common with your cat or with the bacterium living in your gut, but in fact, in terms of basic biology, there is little to choose between you. Your biochemistry is almost identical. The molecules of your body, because of the way they are made up, possess the ability to rotate polarized light in a leftward direction. This is characteristic of life on Earth – if and when we discover lifeforms on other planets, there is a 50 percent chance that their molecules, unlike ours, may rotate polarized light to the right. You, the cat and the bacterium also share the feature that your body is made up of cells, the boundary between one cell and the next consisting of a very definite barrier called a membrane.

Just in case you were thinking that there was nevertheless a basic difference between you and that bacterium (not to mention the cat), there are two types of important complicated molecules shared by all of our planet's lifeforms: nucleic acids and proteins.

For these reasons it seems reasonable to suggest that all lifeforms on Earth shared a common origin: there is little likelihood that it sprang up contemporaneously in several different parts of the globe. However, there is something of a mystery as to what caused that origin.

One idea that has been suggested is that the precursors of life – complex organic molecules – arrived here from outer space. Supporters of such hypotheses fall into three main camps. First there are those theorists, notably Fred Hoyle and Chandra Wickramasinghe, who propose that the interiors of comets are much more likely sites for the formation of organic chemicals than the surfaces of planets. It is almost certain that, during the Earth's early history, the planet was a frequent victim of cometary impacts. When this happened, say Hoyle and Wickramasinghe, the organic molecules within the comets were spewed out over the land. Further chemical reactions led to living organisms.

A second idea is known as the 'bootstraps' theory. The proposition is that, some billions of years ago, visiting extraterrestrials arrived in their spaceship to view the Earth. In one way or another they 'infected' the youthful planet with living organisms – perhaps, before their departure, they emptied out their chemical toilets, leaving a thriving population of bacteria. Of course, there are two basic problems here:

- there is no possible proof of the hypothesis
- the hypothesis may seem to give a palatable explanation of the emergence of life on Earth, but it merely pushes the problem one stage back: how did life originate on the planet of these putative alien space travellers?

A third possibility connected with the idea that life on Earth originated from space is manifestly a nonsense – except for the fact that some remarkably distinguished scientists have supported it, the most recent of them being Francis Crick, winner with James Watson and Maurice Wilkins of the 1962 Nobel Prize in Physiology or Medicine for their work in unravelling the structure of DNA, the 'double helix'. In his 1981 book *Life Itself* Crick suggests that life came to our world because a billennia-old extraterrestrial civilization deliberately sent out vast numbers of spacecraft packed with living spores, hoping thereby to 'seed' countless potentially life-

ABOVE *A knotted wrack of algae. The algae are among the simplest of plants and appeared very early in the history of life on Earth.*
RIGHT *The volcanic island Surtsey, off Iceland, emerged from the sea in 1964. Volcanic eruptions like this one probably assisted the development of life on our planet.*

supporting planets in the Galaxy. This theory is, of course, subject to the two criticisms mentioned above. However, it is worth a little further attention because it is a recent example of a whole group of theories that come under the umbrella heading of *panspermia*. (Both the Hoyle/Wickramasinghe proposal and the 'bootstraps' theory can likewise be considered as variants of the panspermic hypothesis.)

The word 'panspermia' in itself gives the clue to the common element in this group of theories. The Greek prefix *pan* means 'all'; the *spermia* part of the word comes from the Greek for 'seed'. 'Panspermia', as a word, therefore refers to those theories which insist that life came to Earth because 'spores' were somehow blasted off into space in the hope that they might eventually encounter a life-supporting planet – like Earth.

The first panspermic hypothesis can be traced back to 1743 and Benôit de Maillet, who suggested that the germs of life came to Earth from space; they fell into the oceans and in due course grew into fish and, later, amphibians, reptiles and mammals. Much later, in 1908, the first rigorous formulation of the panspermic hypothesis appeared: this was produced by the Swedish chemist Svante Arrhenius (1859–1927). According to Arrhenius, the Universe was packed with 'proto-life' spores that drifted between the stars, their interstellar voyages powered by the radiation-pressure of light.

At the time, the proposal was plausible although, of course, it still ducked the issue of where the spores had come from in the first place. Bacterial spores can withstand extremes of hot and cold, not to mention the vacuum of space, for extended periods, germinating only when they attain more favourable environments. What Arrhenius did not know was that there is a large amount of X-radiation washing about in space, and to this his hypothetical spores would certainly have been vulnerable.

The trouble with all panspermic theories is that they look to outer space, yet we have no particular reason to believe that life could not have emerged right here on our own world. A further difficulty concerns the sheer numbers of such spores that would need to be involved. Imagine that we, the human species, released one billion living spores into the Universe, sending them off in random directions. By the time the spores reached the distance of the 'nearby' star α Centauri, 4.3 lightyears away, the spores would be about 13 million million million kilometres apart from each other.

Bearing in mind that the Earth is only about 150 million kilometres from the Sun, the chances of one of our spores being captured by any potentially life-sustaining planet of α Centauri can be seen to be very slender indeed. Of course, we could scatter many billions of spores, thereby reducing the odds, but even so the possibility of even one of them encountering a suitable receiving planet anywhere in the Universe is vanishingly small.

Here the Crick hypothesis has an advantage over the others, in that he proposes that those mysterious extraterrestrials did not simply scatter spores willy-nilly but sent out a computer-controlled spacecraft containing the spores as cargo. It is easy enough to programme a computer to direct spacecraft towards a long succession of likely stars, to carry out simple tests to establish whether any of the planets of those stars might be capable of sustaining life, and, when appropriate, to dump consignments of spores into the planet's atmosphere. Such a 'guided' enterprise – Crick calls the hypothesis 'Directed Panspermia' – would avoid the colossal wastage involved in a random scattering.

There is, however, a fundamental problem: motive. Interstellar spacecraft, and the computers to control them, do not come cheap. However advanced our hypothetical extraterrestrial civilization might be, building just one of these spacecraft would represent a significant investment. It might take millions of years before the spacecraft encountered a suitable planet, and billions before the result of its efforts would produce an intelligent lifeform, if at all. Taking all this into account, we have to ask why the extraterrestrials should be remotely interested in 'seeding' distant planets. To put it more simply, why should they bother?

Let us return to more conventional theories.

Our planet formed about 4.5–5 billion years ago. It should not be imagined, however, that the newborn Earth remotely resembled the world in which we live today. This was a very hot body, spewing its heat and gases copiously into space: it can be thought of as a planet-sized erupting volcano. In a comparatively short period – a few hundred million years or so – the Earth cooled down, and by this time, because of the gases belched out by its numerous volcanoes, it had an atmosphere. This atmosphere was not especially hospitable:

RIGHT *Eruption of Halemaumau, Hawaii, in 1961.*

LEFT Eruption of Kilauea Iki, Hawaii, in 1959. BELOW Fountains of fire paint the sky in a 1973 eruption of Eldjfell, Iceland. RIGHT A NASA artist's impression of the surface of the planet Venus, a place which has frequently been described as the true-life equivalent of Hell. Volcanism on Venus might assist the development of life there were it not for the devastatingly high surface temperatures.

most of its gases were extremely noxious. However, it is believed that in the virgin planet's oceans there evolved, by chance, complicated molecules called 'proteinoid globules' (or 'proteinoid microspheres'); alternatively, these molecules might have formed on the slopes of the primordial volcanoes and been washed, by rain, down to the seas. Various experiments done in volcanic environments have shown that proteinoid globules are very likely to form in such regions. Another possibility is that the globules could first have been created in the atmosphere by the electrical activity of thunderstorms. In all cases, it appears that the transition from 'complicated molecule' to life occurred in the oceans.

Protein molecules are the precursors of life. If you stir enough of them together and stand back for a few hundred million years, the result is likely to be a chemical entity capable of reproducing itself – one of the fundamental differences between living and nonliving material. This is what seems to have happened early in the history of the Earth, because the oldest known fossil remains of proteinoid globules are believed to date back about 4 billion years. The first living organism cannot have followed far behind.

If these theories are by and large correct we should expect to find proteinoid globules occurring spontaneously in modern volcanic environments. There is some evidence that this is indeed the case, but it is very scant. One reason might be that, as soon as a conglomeration of chemicals becomes complicated enough to become 'protolife', something comes along and eats it. This is not the flip comment it might seem. The notion is that, in countless regions of the world, life is at this very moment coming into existence – almost. We are the descendants of the first wave of life-formation; all others have been devoured by members of our 'generation'.

It is inherent in such ideas that life must be fairly common throughout the Universe. As yet we have no convincing evidence that this is the case.

We have talked loosely about the original atmosphere of the Earth, and the way in which it affected and perhaps still affects life. It is time to turn the tables and consider how the existence of life may have affected the atmosphere of the early Earth.

The primary gases emitted by active volcanoes are carbon dioxide, water vapour and compounds of nitrogen.

There is no reason to believe that this was not true also of the world's earliest volcanoes. Yet the Earth's atmosphere currently consists of about 78 percent nitrogen, 20 percent oxygen and 2 percent other gases. (The proportion of water vapour varies.) Something must have happened to convert the original stew of gases into the ones we breathe today.

According to current theories, all of this is entirely predictable, depending solely upon the distance the Earth is from the Sun. For the sake of comparison we can look at the planet Venus, which is much closer to the Sun than the Earth is, and the planet Mars, which is further away. From the very start the surface temperature on Venus must have been much hotter than that on Earth – purely because of the planet's proximity to the Sun. Most of the water vapour emerging from Venus's early volcanoes would therefore remain in the planet's atmosphere, in the form of clouds. The longer this persisted, the hotter the planet got, because energy from the Sun could penetrate down through the clouds but was incapable of escaping back out through them (the famous 'greenhouse effect'). In due course the surface temperature on Venus was higher than the boiling point of

water, so that there was no longer any possibility of liquid water on the planet's surface – in other words, there could be no rivers or seas. Mars, by contrast, being so much further away from the Sun, is very cold: the water vapour produced by its early volcanoes would swiftly have frozen solid, leaving a thin atmosphere composed largely of carbon dioxide. (There is evidence, though, that once upon a time Mars enjoyed running water.)

By contrast with both of these other planets, the Earth was just the right distance from the Sun. At its surface, water could exist as a liquid rather than as a gas, and so oceans and seas came into existence. Water is capable of dissolving carbon dioxide; the composition of the atmosphere would therefore change considerably. Nitrogen, on the other hand, is a very inert (nonreactive) gas: the quantities of it emerging from the volcanoes might have been, in percentage terms, extremely small, yet they stayed in the atmosphere.

However, so far as we are concerned the important constituent of the atmosphere is oxygen: without it we would never have come into existence. The clue to the plenitude of oxygen in the Earth's atmosphere would appear to be the respiratory activities of the lowest and earliest forms of terrestrial life. These, like plants do today, took carbon dioxide from the atmosphere, stripped away the carbon and emitted oxygen in return. We can date the 'oxygen revolution' – the time when oxygen became an important gas in the Earth's atmosphere – fairly precisely: it occurred about 410 million years ago. This can be deduced from the fact that rocks of this age are typically red: the iron in them rusted, a process which cannot occur without the presence of oxygen and water.

As yet, no one knows exactly how life originated on Earth. The possibilities seem almost endless. The seeds of life might have appeared spontaneously here or in space, or they might have been deliberately sent here by an extraterrestrial intelligence. The latter possibility is, to be polite, unlikely; the other two imply that life is very widespread throughout the Universe. If so, why is that we have not heard anything from all those other civilizations out there? We shall return to this question later.

LEFT *The surface of Mars as photographed in 1976 by Viking 2.*
BELOW *Old Red Sandstone*

WHY DID THE DINOSAURS DISAPPEAR?

At the moment there is a single dominant species on our planet, *Homo sapiens* – in other words, you. But about 65 million years ago the picture was very different. The dominant animals were reptiles. Their size varied considerably, from behemoths like *Brontosaurus* to little scuttling beasts about the size of a hen.

The dinosaurs died out very suddenly. The reasons for this extinction are not well understood. It seems almost certain that a large comet hit the Earth, throwing up vast quantities of junk into the air so that the planet's surface temperature dropped drastically. Geological evidence supports this theory. At the same time it does not seem enough. Our own species, if similarly threatened, would probably survive – not a lot of us, but enough to continue the species.

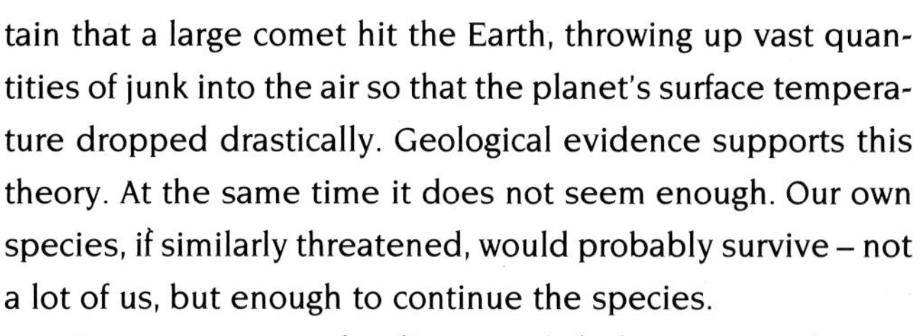

For some reason the dinosaurs failed to survive: this is a considerable mystery. At the time of their demise the land

RIGHT *Skeleton of* Triceratops
BELOW *Building on such skeletons, artists can come up with impressions of what dinosaurs like* Triceratops *looked like, but it should be stressed that such images are no more than guesses.*

LEFT Barringer meteorite crater, between Flagstaff and Winslow, Arizona. The crater is about 1.280km across and about 175m deep. FAR RIGHT *A fossil of* Archaeopteryx, *believed to be the ancestor of modern birds. Recently some scientists, including Professor* Sir Fred *Hoyle, have claimed that these fossils are forgeries.* BELOW RIGHT *A large female crocodile on the* Nile River. *Why the crocodiles should have survived the extinction of dinosaurs is a mystery.*

was infested by the mammal-like reptiles, which had the looks and probably the habits of rats: we are the descendants of such creatures. One current explanation as to why the dinosaurs died and the mammal-like reptiles survived is that the dinosaurs were cold-blooded: when conditions became arctic, the dinosaurs just died. Modern theories, however, suggest that the dinosaurs were warm-blooded, like the mammals (including ourselves) that currently dominate the Earth. When the dinosaurs died, why did the mammal-like reptiles not do likewise?

The short answer to that question is that nobody knows. As noted, it seems very likely that, 65 million years ago, a comet did indeed hit the Earth. The impact threw up vast quantities of material into the upper atmosphere, where it stayed for months or years, blotting out the light of the Sun; some of this material came from the comet, some from the surface of the Earth. The net result was that much of the energy from the Sun was reflected back out into space, so that the surface of the Earth became much cooler than it had been before. Also, without sunlight, plants cannot photosynthesize, so there must have been a very considerable food shortage. In fact, the effect on all forms of life may well have been as devastating as the 'nuclear winter' which will occur should the nations of Earth ever engage in a nuclear war. To extend the parallel, the impact of the comet would also have released large quantities of hard radiation.

In a certain way, of course, the dinosaurs were not really exterminated: we see their descendants all around us, in the form of modern reptiles and birds. Looked at from this angle, the mystery may not be such a mystery after all. We can imagine the following series of events:

- animal life evolved until the reptiles dominated all other forms
- the impact of the comet obliterated large portions of the Animal Kingdom, with mammals suffering just as severe depredations as the reptiles (fishes and other marine animals probably came out of it all comparatively lightly)
- second time around, by chance or otherwise, as animals further evolved to repopulate the Earth, the mammals came out on top (although a case can be made that it was really the insects that came out on top!)

A further clue could be the sheer size of the dinosaurs. When there is a severe food shortage, the first to suffer are the larger animals. The giant herbivorous dinosaurs would have been the first to go, as plant matter became scarce, but they would soon be followed by the large carnivores that had preyed upon them. Naturally, these predators must have turned their attentions to other animals, like the mammal-like reptiles; but we must remember that these animals were *small*. To keep a *Tyrannosaurus rex* alive would require catching and killing a tremendous number of mammal-like reptiles. Add to the beast's problems that it was already enfeebled by lack of nutrition, and one can see that its chances of survival were really pretty low. In other words, when the disaster came, it was the big animals that were most vulnerable, and at the time the big animals were all, as it happened, dinosaurs.

We can note in this context that the big animals of today, the whale and the elephant, are quite recent arrivals. The earliest known elephants date from about 50 million years ago – well after the event – and were then only about the size of a pig. The earliest known whales date from about the same time, but in this case the situation is slightly more complicated, in that these early whales were very similar to modern ones, and could be big – up to 15m (50ft) long. By comparison, *Diplodocus*, the biggest of the dinosaurs, was typically 28m (91ft) long, while a modern blue whale can reach a length of about 30m (100ft). Nevertheless, a 15m (50ft) animal is no mean beast; moreover, because of the way that the earliest known whales so closely resemble modern ones, we have to assume that their ancestry stretched back some way, even though no relevant fossils have yet been

discovered. Since 15 million years is a very short period in terms of evolution, we have to conclude that there were some big whales around at the time of the catastrophe. However, as noted in passing in the case of fishes, conditions in the oceans are likely to have been less drastically affected than those on land.

Another puzzle is that one kind of large reptile survived – the crocodiles. Some of these animals were and are big: one species which existed around the time of the catastrophe was 15m (50ft) long. Of course, these animals are semi-aquatic, but much of their prey is terrestrial. One can only assume that, at least for a while, they were able to survive on fully aquatic prey, such as fishes. Besides, there is good evidence that the crocodiles of that time were much faster-moving reptiles than their dinosaur contemporaries – indeed, the ancestors of modern crocodiles may have been nimble land-animals, since to this day crocodiles have an ankle structure typical of a rapid terrestrial carnivore. Perhaps the crocodiles, through their speed, were able to kill sufficient numbers of smaller animals to survive the 'winter'. More likely, they sustained themselves by a mixture of both strategems.

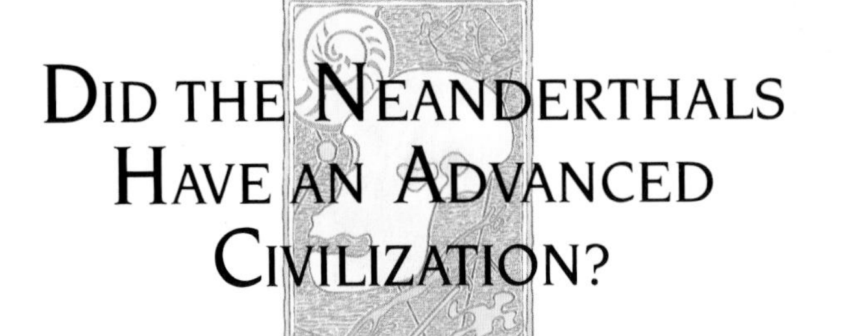

Did the Neanderthals Have an Advanced Civilization?

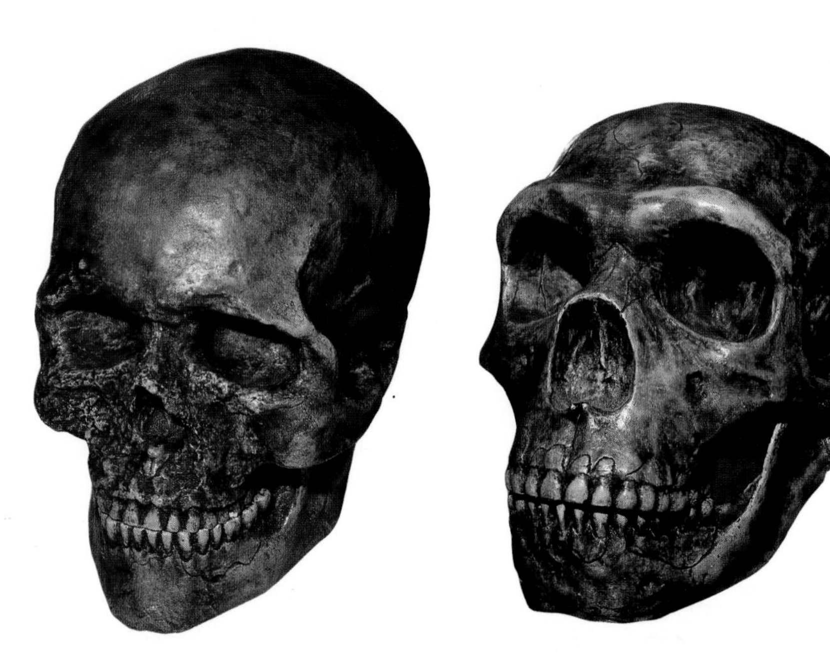

FAR LEFT *Reconstruction of a Cro-Magnon skull.* **LEFT** *Reconstruction of a Neanderthal skull.* **FAR RIGHT** *Dugout canoes in Nepal. Such craft have been in use since humanity's very earliest days; they are an example of an invention which serves its purpose so well that there is little need to improve upon it.*

The idea of any form of prehistoric human beings having a civilization that we could reasonably described as 'advanced' seems to be a contradiction in terms. After all, one aspect of civilization is the development of a written means of communication, which implies that records are preserved. When this happens, there is no longer any such thing as prehistory: it has been replaced by history.

However, it has been pointed out – most recently by Stan Gooch in his *Cities of Dreams* (1989) – that this is to take a rather narrow view of what we mean by civilization. Because of our own cultural heritage, we think in terms of things like the construction of permanent artefacts and the invention of useful tools like writing and mathematics – in other words we are, essentially, technologically oriented. But is this the only way in which a complicated and highly structured civilization can evolve?

The answer has to be a resounding 'no'. We can legitimately think of an ants' nest as a civilization: its members carry out highly specific or even very generalized tasks in pursuit of the common good. Yet ants have yet to develop writing, and their major artefacts – their nests – can be demolished in minutes using a spade. Dolphins are highly intelligent beings (they may be as intelligent as we are, and perhaps even more so), yet they have no writing and create no artefacts. Nevertheless, in many dolphin communities the interactions between the various members are commonly more civilized than those current in parts of our cities. There are even human communities around the world which exist without such tools as writing, while constructing artefacts, commonly of wood, that cannot be described as permanent (in terms of thousands or millions of years).

Ants can create a structured society without being possessed, individually, of very much by way of brain. Bees do much the same. One can look at other animals – baboons, chimps and so on – and see a lesser form of the same structure being created. Inevitably we have to conclude that social action, commonly considered a by-product of intelligence, can in fact exist quite independently of that quality.

Neanderthals possessed bigger brains than we do; the Cro-Magnons, generally regarded as having wiped out the Neanderthal precursors, were similarly large-brained. This is not proof in itself that our ancestors were more intelligent than we are: it seems that the ability of a brain depends less on its size than on the extent of its complexity, as epitomized by the degree of crenulation (folding) on its exterior. Unfortunately, since the brains of dead people rot swiftly, we have little notion of the degree of crenulation of the Neanderthal or Cro-Magnon brain.

It is therefore plausible that Neanderthals enjoyed a civilization which, while very different from our own, had its own structures, conventions and complexities. Gooch, in *Cities of Dreams*, postulates a society based on what we can call 'mental artefacts'; that is, ideas rather than physical constructions. The only trouble with such a hypothesis is that it is difficult to prove. Gooch produces evidence that our own society, which he claims is derived from that of the Cro-Magnons, still shows traces of the social mores he ascribes to the Neanderthals. His theory is, essentially, that the Cro-Magnons, in wiping out the Neanderthals, accidentally incorporated the latter's 'dream' ideas into their culture – and that such ideas still survive today.

The hypothesis cannot be dismissed out of hand. Far, far more recently, at around the time when the Roman Empire was at its height, the Celts across Europe enjoyed a 'confederation' of which signs can still be seen today. A pure Scot arriving in New York, Cardiff or any other of the non-Scottish capitals can expect to be welcomed by other Celts, whatever their original nation and however many generations have passed since their ancestors left it.

The theory's great difficulty is, as noted, its considerable lack of proof. One reviewer, John Naughton in *The Observer*, went a little over the top when he remarked: 'How does Mr Gooch know all this? The answer is that he is guessing.' This is unfair criticism – Gooch produces a mass of indicative, albeit necessarily speculative, evidence in favour of his hypothesis – but nevertheless it is true that our knowledge of the past is and has to be based on the artefacts surviving from that past: lacking the artefacts, all we can do is make an educated deduction.

Whether or not we are prepared to accept that the Neanderthals had an advanced but abstract civilization, we are left with a question: what sort of civilization *did* the Neanderthals have? We do not know. We have evidence of cannibalism and other habits that we might consider a trifle unsavoury, but this does not mean that Neanderthal society was necessarily uncivilized. Killing people simply in order to satisfy hunger is hardly one of humanity's more endearing habits (and anyway seems always to have been rare), but the consumption of dead heroes can be seen as a mark of respect, in that the eaters may hope to inherit some of the admired qualities of the deceased. It is possible that wise Neanderthal men and women were only too delighted to allow themselves to be killed and eaten, on the basis that their souls would live on in newly inhabited bodies.

There are a few further questions:

- what sort of a civilization did the Cro-Magnons enjoy?
- did the Cro-Magnons indeed deliberately wipe out the Neanderthals?
- why were the Neanderthals, who as a species of human being had had a much longer pedigree, vulnerable to the Cro-Magnons?
- why, in any case, should the Cro-Magnons have felt the slightest urge to exterminate the Neanderthals (with whom it is thought they could have interbred)?

This last question is probably the most profound. In our own time it has been only fascists, racists and their ilk who have felt the urge to slaughter vast numbers of people whom they have regarded as 'different' from themselves. Thankfully it is comparatively rare for human beings to seek to exterminate those creatures whom they regard as very much their inferiors – it is hard to find someone who favours the meaningless slaughter of dogs, for example. But people are much more likely – and we should stress that only the lunatic fringe is involved – to believe it necessary that those they perceive as only *minimally* inferior to themselves should be annihilated. (The next stage, paradoxically, is to describe the 'inferiors' as animals, so that the slaughterers need have no qualms of conscience. This is how, for example, Hitler persuaded numerous Germans to murder Jews.)

Here we have a possible explanation of why the Cro-Magnons might have exterminated the Neanderthals: the Neanderthals were very like them, and therefore constituted a threat. The Neanderthals were, however, physically weaker, and so it was 'reasonable' to wipe them out, or at least to subjugate them to the extent that eventually they died out. (We see the same sort of philosophy working in countries like South Africa today.)

As to whether the Neanderthals had an advanced civilization, there is no possible answer to this question. Any traces have been obliterated by the advance of the ethos that we have inherited from our Cro-Magnon ancestors. We are now a technological species, and until comparatively recently we paid little attention to species that were not.

DID EXTRATERRESTRIALS VISIT THE EARTH IN THE DISTANT PAST?

There is a succinct and spiritually satisfying answer to this question: probably not.

There are several reasons for the response, of which two are significant. The first is that evidence for such a visit seems to be nonexistent. The second – much less scientific – is that the arguments put forward in popular potboilers supporting such a notion are in general so specious that one is embarrassed to allow oneself to be associated with them. It is largely for this latter, purely emotional reason that very few scientists will take the idea seriously. Those who do keep very quiet about it, although, as we have seen, the 'bootstraps' hypothesis concerning the origin of life is never totally dismissed by any but the most hidebound of life scientists.

The ancient-astronaut industry really got under way in 1969, with the first English-language publication of a book called *Chariots of the Gods?*, by Erich von Däniken. Von Däniken's first few lines – 'This was a very difficult book to write. It will be a very difficult book to read' – were true but less than prophetic, since people read it in their millions.

In fact, there were earlier ancient-astronaut theorists, including Desmond Leslie, who wrote with George Adamski the notorious *Flying Saucers Have Landed* (1953). Leslie tells us that aliens from Venus arrived on Earth in the year 18,617,841BC. He knows this because he has been able to decipher 'ancient Brahmin tables'. This comes as something of a surprise, because the human species is only about four million years old, and the Brahmins substantially younger!

LEFT *George Adamski, the most famous of all claimed UFO abductees, peering through a Newtonian telescope . . . during the day!*

Another early theorist was W Raymond Drake, whose 1964 book *Gods or Spacemen* is heartily recommended to those in search of innocent mirth. He is, perhaps, in with a chance when he equates gods with the planets that bear their names: ancient legends of Jupiter doing nasty things to people could really represent an ancient memory that aliens from the *planet* Jupiter dropped nuclear bombs on Earth. All well and good until he extends his ideas to the god Uranus, which he enthusiastically does. Unfortunately, the planet Uranus was not discovered until 1781.

In a book published in 1974, *Mystery of the Ancients*, Craig and Eric Umland claimed that the aliens were still among us, in the form of the Maya. According to the Umlands, scientists in the USA and the USSR are currently competing to decipher the secrets of Mayan script in order to unlock the secrets of the Universe. Besides, the Maya knew of the wheel – and how could they have done so unless they had been told about it by extraterrestrial visitors?

All of these ideas are fun, of course, but they do not help us to answer the original question. If extraterrestrials visited humanity in the very early days of our species, the cultural impact must have been major – unless, that is, the extraterrestrials had the common sense not to interfere too much with whatever level of civilization then existed.

Very few studies have shown any evidence that extraterrestrials have ever visited the Earth. As always, however, there is an exception. Robert Temple, in his *The Sirius Mystery* (1976), examined reports of anthropologists concerning the Dogon tribe of North Africa. His analyses suggested that the Dogon knew that the planets of the Solar System went around the Sun in ellipses, rather than in circles; they knew also that the dwarf-star companion of Sirius (the Pup) existed and was made up of very compressed matter. The Dogon were aware that the planet Saturn had rings and that the planet Jupiter had four moons, which are pieces of information that cannot be obtained without the use of a telescope – an artefact the Dogon have yet to discover.

Here we have a quandary. Jupiter has many more moons than four – we are still not certain of the number, and possibly never will be. The rings of Saturn are indeed spectacular, but we now know that Jupiter, Uranus and Neptune likewise have rings: it is strange that the visiting extraterrestrials should not have mentioned this.

The Dogon were almost certainly visited late in the nineteenth century by missionaries who were up-to-date in terms of modern science. The small, hot and compact companion of Sirius was discovered in 1915 – the first real investigation of the Dogon did not take place until at least the 1930s and 1940s. But it seems far from unlikely that other explorers might have visited the Dogon in the intervening period, passing on the latest scientific information.

There is another aspect to the ancient-astronaut hypothesis: perhaps our remote ancestors could not have survived had it not been for the fact that they were *destined* to do so. But such ideas are heresy . . .

Has the Earth Flipped Over at Least Once in the Past?

In 1982 Peter Warlow published a book called *The Reversing Earth*. It is generally recognized that the magnetic field of the Earth has periodically reversed: Warlow claimed that the Earth as a whole had flipped over. In so doing he was supporting the ideas of Immanuel Velikovsky (1895–1979), one of the earliest of the modern breed of pseudoscientists.

Velikovskianism seems set to make a sudden resurgence after all these years. The neo-Velikovskians are, refreshingly, starting to apply some of the tools of science in their arguments: such knotty subjects as physics and mathematics are beginning to make their appearance. Unfortunately, they are doing so alongside all the inherited pseudoscience and woolly thinking, the pretentious massing of obscure and unreliable (and unreliably dated) data which have traditionally marked Velokovskianism off from more orthodox science.

Velikovsky's ideas are wide-ranging, but we can sum them up by saying that, about 3500 years ago, the planet Venus was born as a comet, spat from a volcano on the planet Jupiter. This comet lurched around the Solar System before settling into its current orbital position. Most notably, it had several close encounters with the Earth, causing great upheavals, dividing the Red Sea, spattering the landscape with flaming petroleum, and in general making life sheer hell for our ancestors. One of its major effects on our planet was to cause it to flip 'upside-down.' on its axis.

Velikovsky's *Worlds in Collision* produced an overreaction from the scientific establishment for the simple reason that Velikovsky had cast his net far and wide in search of supporting evidence for his theory. Since no scientist was prepared to sit down and do a comparable amount of research in order to demolish a theory which seemed so patently to be a load of rubbish, they all simply cried 'Rubbish!' and stamped their feet. Astronomers noted that the archaeology was interesting and archaeologists said that the astronomy seemed to make a lot of sense.

The overall effect has been unfortunate in two different ways. First, the Velikovskians have been given the chance to make a superficially plausible claim that there is some kind of 'orthodox-science' vendetta against them and against their 'master': they have been able to portray themselves as persecuted martyrs. Second, the furore obscured the fact that Velikovsky was making an important point: catastrophes *have* occurred in the past. Until recent years this idea was largely pooh-poohed by the scientific community. A second point Velikovsky made was that evidence of such catastrophes could be deduced from ancient writings, oral legends and the like.

There were, of course, flaws in Velikovsky's reasoning: his imagined catastrophes were fanciful and, while it is all very well to derive clues from oral legends, it makes little sense to prefer these to properly researched results.

In *The Reversing Earth* Warlow tried to prove that, during the last 13,000 or so years, the Earth has flipped over on its axis about five times. He employed the conventional Velikovskian technique of selecting evidence while at the same time adopting much of the traditional (that is, false) Velikovskian lore. It is worth going through his arguments in some detail, because they represent the very best of the theories concerning the idea that the Earth has periodically flipped on its axis.

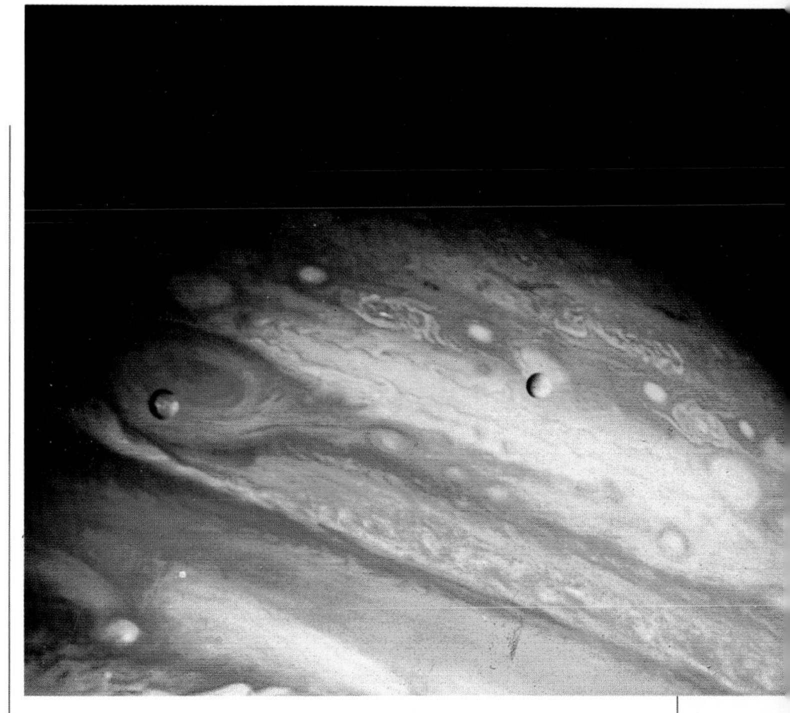

FAR LEFT *The Earth from space, as photographed in 1972 by Apollo 17.* **ABOVE** *Photograph of the giant planet Jupiter taken in 1979 by Voyager I, showing two of the planet's moons, Io and Europa. The Great Red Spot is clearly visible.*

In the ordinary way, turning over a massive spinning object like a planet is a difficult task. Anyone who has ever played with a gyroscope will know that the object displays a distinct aversion to being turned upside-down; in many ways the spinning Earth can be viewed as a sort of super-heavy gyroscope. Warlow pointed out, quite correctly, that this is not the sort of inversion we should be looking at: if the Earth flipped over in this way the Sun would still rise in the east, whereas there are legends that, before the (undescribed) catastrophe, the Sun rose in the west, and that it is only since that catastrophe that the Sun has risen in the east.

There is, however, a different way in which spinning objects can turn over. This is exemplified by a children's toy called the tippe-top. Most tippe-tops come from Christmas crackers; they look like an apple with a fat stem. The tippe-top has two delights. First, if you spin it on the table, 'stalk'-upward, it almost immediately flips itself over to spin upside-down, balancing on the stalk'. The second and more exciting aspect is that, when it does so, the direction of spin is not affected: if the initial spin was clockwise, it will still be clockwise even after the toy has turned over. This is exactly the opposite of what common sense might predict. We can note, too, that the tippe-top differs markedly from the gyroscope: a tippe-top turns over spontaneously – indeed, it is hard to stop it from doing so – whereas a considerable amount of effort is required to flip over a gyroscope.

This, then, would appear to be an easy way to turn the Earth over – and one in better accord with the legends than the (probably impossible) straightforward one. However, problems arise as soon as you start to think about what is actually happening when a tippe-top flips over. The mathematics – and indeed the physics – of the situation are extremely complicated, but we can note that the tippe-top is not operating in isolation. The complete system involved includes a flat surface – a table, perhaps – and a steady downward gravitational pull. Unfortunately, there aren't any tables in space.

What, even so, of the required gravitational pull? To answer this question we are led back to the idea of big gassy planets like Jupiter spitting out small rocky ones like Venus, which then play a sort of cosmic billiards before settling down. Warlow and his supporters point to the fact that certain quasars (highly energetic galaxies) have jets of material emerging from their cores: could it not be that stars and big gassy planets can behave analogously? This is only superficially a valid question, since quasars are in every respect very different objects from stars and planets. Also, the jets of material associated with them seem certainly to shoot out from the rotational poles and to keep travelling that way; were new planets to engage in the game of cosmic billiards they would have to shoot out equatorially from their 'parents'.

There are further problems with the whole scenario. Jupiter is a very massive planet, and its escape velocity is correspondingly high. Venus is much smaller, but nevertheless still has about the same mass as the Earth. To accelerate a body with the mass of Venus to the escape velocity of Jupiter would require a fairly impressive 'spit'!

Then again, for a planet with Venus's mass to have a gravitational effect on the Earth even as strong as that of the Moon, it would have to pass within about four million kilometres (2,500,000 miles). This distance may seem large, but it is dwarfed by the length of the Earth's orbit – 9,300,000,000km (5,800,000,000 miles). In other words, the chances of Venus having even this negligible effect on the Earth are only about 1 in 115 per transit of our planet's orbit. And this is assuming that Venus is travelling in exactly the same plane as the Earth's orbit, which it probably wouldn't be. If we allow Venus five transits of the orbit, the chances of its affecting the Earth are still only about 28:1 against.

Warlow's claim that the Earth has flipped five times over 13,000 years implies a cosmic near miss every 2600 years. So where are all the planets? Even laundering the figures in Warlow's favour, only one out of every 28 newborn planets will have any effect on us, so planets must be popping out from somewhere every century or so. There must be tens of millions of them out there!

One can go on in this knockabout vein for some time, because some of the illogicalities of the Velikovskian scientists are very funny indeed. However, this does not mean that they are necessarily wrong in their idea that the Earth may have flipped over, simply that they haven't adequately worked out what could cause it to do so. And we have to acknowledge that, although it is difficult to flip planets over, there is good evidence that similar things have happened in the past among the outer planets of the Solar System. For example, Uranus, which is much more massive than the Earth, seems to have been tipped over so that it lies on its side. Triton, the largest satellite of Neptune, goes round the

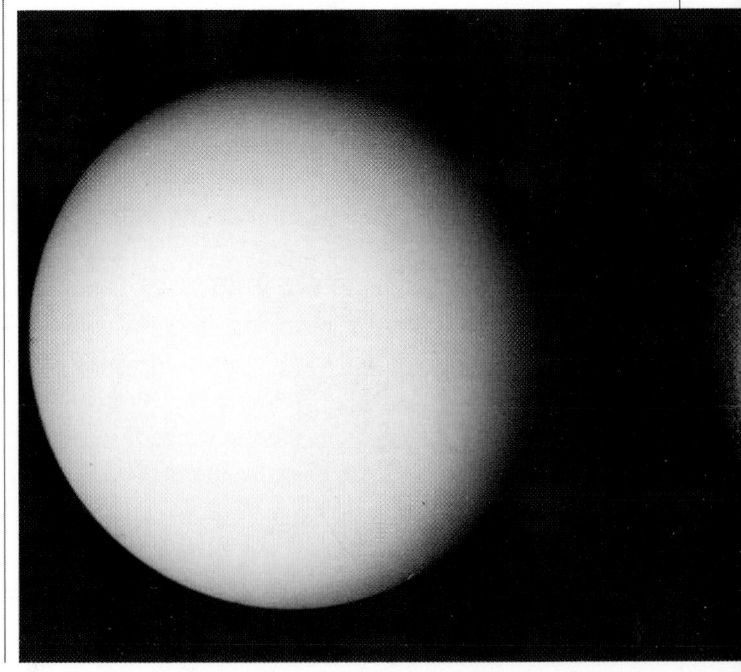

planet in the opposite direction to Neptune's other moons, and may originally have been a small planet in its own right which was captured by Neptune's gravity. The little double planet Pluto has such a weird orbit that, even though almost all of the time it is by a long way the furthest known planet from the Sun, occasionally (as at the time of writing) it is closer to the Sun than is Neptune. One explanation for this is that the two bodies constituting Pluto were originally moons of Neptune, but that something tore them from the giant planet's grasp – the converse of what is thought to have happened to Triton! Clearly some very large forces have disrupted the outer regions of the Solar System at some stage, or stages, in the past, but we have no idea of when or of what they might have been.

So it is not an impossibility that the Earth has indeed been flipped over, or at least tilted, during remote or even not so remote prehistory.

Before leaving this particular mystery, we should repeat that there is good proof that the direction of the Earth's magnetic field has indeed flipped many times during our planet's lifetime. (The reasons are fairly well understood, but need not concern us here.) Interestingly, the next flip is rather overdue, so do not be surprised to wake up tomorrow and find that your compass isn't working properly!

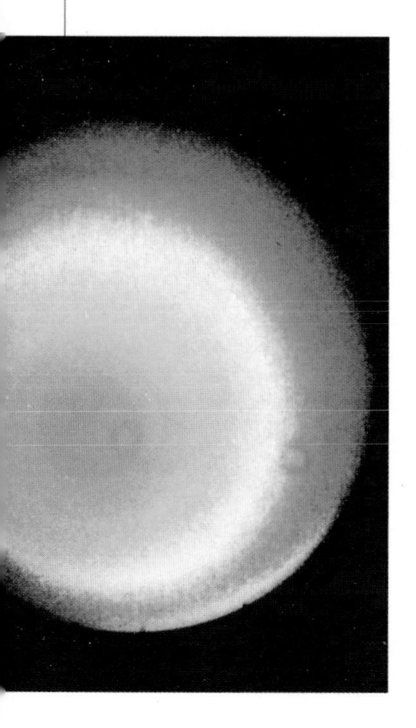

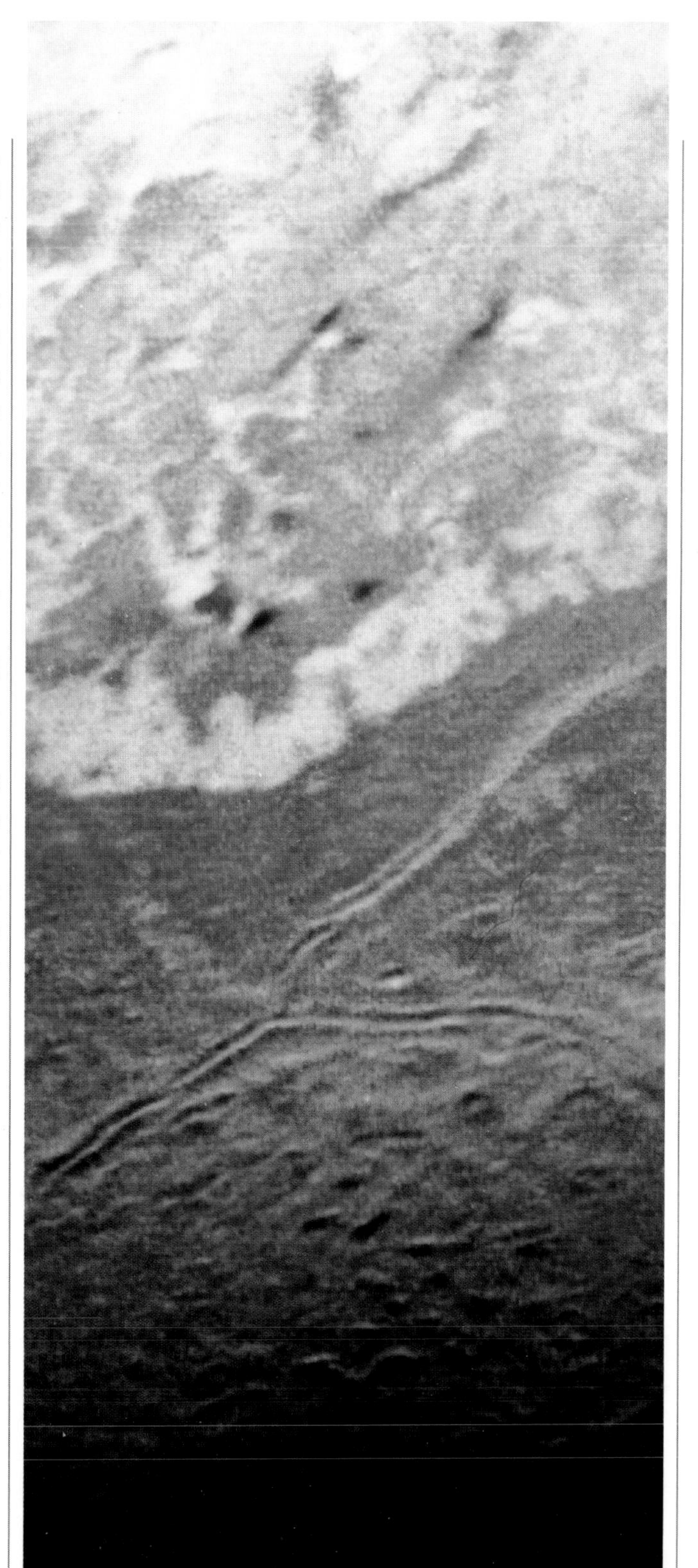

LEFT *Images of Uranus built up from information sent back in 1986 by Voyager 2; the image on the right is in false colour and extreme contrast enhancement to show details of Uranus's polar regions.* **RIGHT** *A Voyager 2 image from 1989 of part of the surface of Triton, Neptune's large moon.*

Is There Any Truth in Tales of Lost Continents?

Ah – Atlantis, Lemuria, Mu, Avalon . . . Where are you now? More to the point, did you ever exist?

The term 'lost continents' invokes two quite separate phenomena, which are often muddled through ignorance or deceit by those seeking to 'prove' the one-time existence of Atlantis and the rest. So we should establish at the outset that, over geological time, certain continents have indeed been 'lost'. This is because of the process known as plate tectonics.

The surface of the Earth is made up a number of plates, and these move relative to each other. The continents and other landmasses perch on top of the plates, and consequently likewise move relative to each other. The plates are not permanent: they are constantly being replenished by liquid rock gushing up from the Earth's mantle (the subsurface layer) at the long submarine mountain ranges known as the mid-ocean ridges. Clearly, if a plate is gathering new material along one margin, it must lose material at another. (There are a few theories that suggest that the Earth as a whole is expanding, but none are very convincing.) At what are called 'destructive' plate margins, the surface material of one plate is subducted (that is, thrust down) beneath the surface material of the adjacent plate, rejoining the molten rock of the mantle. This picture is complicated should one or, especially, both of the plates be carrying a continent: the two continents collide with all the subtlety of a road accident. The continental rock buckles up to form a mountain range. The Himalayas, which resulted from the collision of India with the rest of Asia several million years ago, is one such mountain range.

In this sense, then, there have obviously been continents that have been lost – either subducted back into the mantle or fused onto others. However, the plates move very slowly, so the continental collisions we are discussing happened millions of years before *Homo sapiens* came on the scene. By contrast, the legends concerning Atlantis and the others refer to a period of time only a few thousand years ago. There can therefore be no connection between these legends and the process of plate tectonics.

When we look at the case in favour of lost continents, it seems to make sense to examine Atlantis first. After all, if we could prove the case for Atlantis, the cases for Lemuria, Mu and the others would seem much more plausible. So what do we know about Atlantis?

The primary legend of Atlantis seems to have started in the fourth century BC, when Plato wrote that Solon, who had lived a century or so earlier, had visited Egypt and there had

FAR LEFT *Damage in the wake of an earthquake in Anchorage, Alaska, in 1964. There was a horizontal movement of 4.25m and a vertical movement of 3.35m.* **RIGHT** *Glaciation in the Swiss Alps, with the Matterhorn rearing proudly in the background.*

been told by a priest that, once upon a time, there had existed a continent beyond the Pillars of Hercules (that is, the Strait of Gibraltar). This continent bore a very advanced civilization, but was devoured by the ocean in some unspecified catastrophe. Ever since then – and especially since the US politician and writer Ignatius Donnelly published his *Atlantis, the Antediluvian World* in 1882 – the legend of Atlantis has captured the public imagination.

But is it any more than a legend? Possibly so. Sometime after 2500BC there arose on the Mediterranean island of Crete a civilization that was, for the time, extremely advanced. Today, it is known as the Minoan civilization, after the legends about King Minos, Theseus and the Minotaur. The civilization flourished for more than a millennium until, about 1400BC, the comparatively nearby volcanic island of Thera erupted in a fashion that makes the 1883 Krakatoa event look like a mere squib. A colossal tsunami (or 'tidal wave') destroyed the Minoan civilization in minutes.

The theory is, then, that tales of Atlantis are reflections of this disaster. As the story was passed from lip to lip, the level of destruction became greater and greater, as did the area of the land destroyed. Soon it must have become obvious to the story-tellers that such a huge land could no longer be fitted into the Mediterranean, and so it was resited beyond the Gates of Hercules.

What, then, of the other 'lost continents'?

Lemuria, normally located in the Indian Ocean, seems first to have been proposed by the English geologist Philip Lutley Sclater (1829–1913). He suggested that such a continent might have existed because he could see no other possible explanation for the modern distribution of lemurs. The German naturalist Ernst Heinrich Haeckel came out with a similar theory. Another supporter was Alfred Russel Wallace (1823–1913), co-author with Charles Darwin (1809–82) of the theory of evolution by natural selection. Sadly, there seems little reason to support the notion.

Mu is on even shakier ground. This continent seems to have been created by Augustus Le Plongeon for his 1896 novel *Queen Moo and the Egyptian Sphynx*, which he claimed was based on original Mayan writings – which writings, of course, just happened to be unavailable to other scholars. Some while later Colonel James Churchward picked up the idea of Mu. He visited various Indian or Tibetan (there is some confusion here) monasteries and discovered a host of stone tablets on which were carved messages in an unknown language. It took Churchward a while to realize that this language was in fact Naacal, the first language of humankind, which could be translated by, at best, a couple of Indian (or maybe they were Tibetan) sages. Thanks to this act, Churchward was able to tell the world that Mu sank into the chilly waters of the Pacific 12,000 years ago, that almost all of its 64 million inhabitants perished, that a few survived on Pacific islands, that these survivors gave birth to *Homo sapiens*, that white people are nicer than black because they're closer to the original folk of Mu, and all the other nonsense you might expect. All one can say about the legend of Mu is that it was invented recently and has no scientific backing whatsoever.

One characteristic common to all of the fabled 'lost continents' is that they were ideal places to live. This would suggest that they have the same scientific status as Heaven.

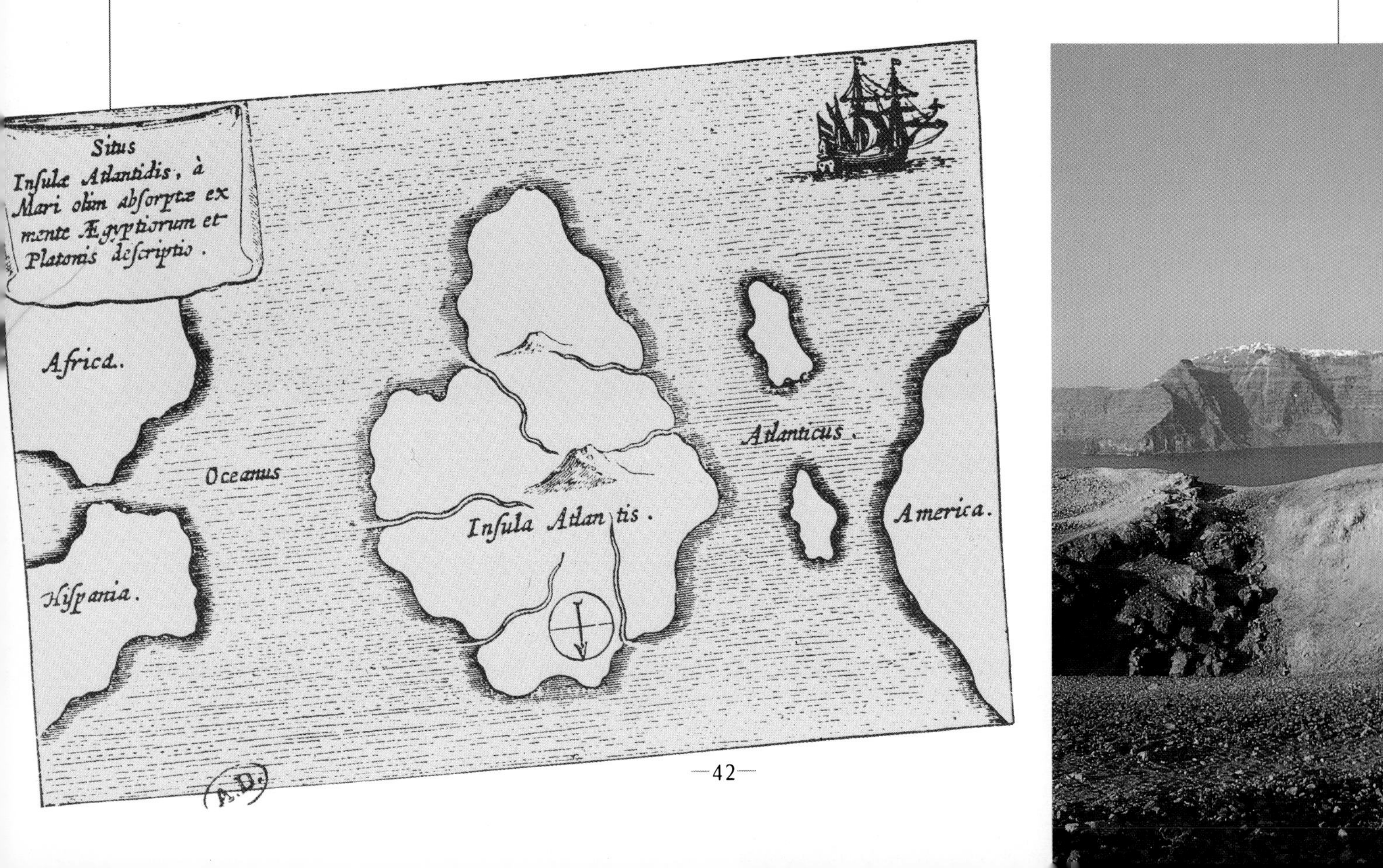

FAR LEFT *Map of Atlantis from Athanasius Kircher's* Mundus Subterraneus *(1644); note that north is at bottom. Kircher's guess as to the continent's position was based on Plato's version.* **LEFT** *Copy of a fresco found at Knossos, Crete, showing bull-dancing.* **BELOW LEFT** *The island of Santorini.* **BELOW** *Minoan houses still standing on Santorini, having been buried in ash during the eruption around 1400BC.*

WHY DO ICE AGES HAPPEN?

One of the most infuriating habits of the scientifically illiterate is to talk of the 'ice age' as if there had been only one. In fact, there have been several ice ages, possibly many. Furthermore, it seems likely that the 'most recent' ice age is in fact still continuing: we are probably enjoying one of the interglacials, or warmer periods, of this ice age.

There is a certain amount of debate as to why ice ages occur. They seem to have happened periodically over the last 900 million years, and may have been doing so since the early days of Earth (the longer ago something happened, the less likely it is that we will have tripped over the evidence). The ice ages we know about are dated as follows:

- 2300 million years ago (probably two or three ice ages, but the dating here is vague)
- 900 million years ago
- 750 million years ago
- 600 million years ago
- 450 million years ago
- 300 million years ago
- current

Intriguingly, there is very little evidence of any widespread glaciation 150 million years ago, during the geological period known as the Jurassic. This lacuna may be of interest in trying to establish the circumstances, or complex of circumstances, required if an ice age is to occur . . . assuming, of course, that the neat spacing of the others is not just a matter of chance.

A good correlation can be seen between the occurrence of an ice age and various other geographical/geological phenomena:

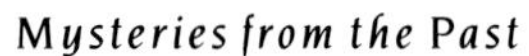

- there is a substantial landmass near one of the poles – in the case of the current ice age, Antarctica fits the bill
- extensive mountain-building, due to colliding continental plates, is in progress – the Andes, Himalayas and Cascades are all mountain ranges that have come into existence during the past 10+ million years, and are still in the process of creation
- more landmasses are above water than usual – although this may be a chicken-and-egg effect because during an ice age much of the world's water is locked up as ice, so that ocean levels are lower

None of these circumstances pertained during the Jurassic, but this does not mean that mountain-building and the proximity of a large landmass to one or other of the poles, taken together, constitute the cause of an ice age. After all, the coincidence could not be expected to occur so neatly every 150 million years.

It is not hard to see how these two phenomena might, as it were, assist an ice age on its way. A continental landmass located at the north or south pole is likely to be ice-covered, and therefore very reflective: a lot of the impinging sunlight will be reflected back out into space. Similarly, high

LEFT *Glaciation in Greenland underscores the fact that the Earth is currently undergoing one of its periodic ice ages. We are lucky enough to be living during a warm period (interglacial) of this ice age, but our luck cannot last forever; at some time during the next few thousand, few hundred or even just few years the next cold period (glacial) will commence unless humanity's pollution contributes so much to the greenhouse effect that global warming prevents the glaciers from spreading. The future of our species seems to be gloomy, either way. However, humanity could survice a glacial; its chances of surviving a runaway greenhouse effect are much more slender.*

RIGHT *The Pleiades cluster. The ancient Greeks identified and named the seven brightest stars; today we can see only six with the naked eye. In fact the cluster contains countless young stars.*

mountains (and mountains are highest when new, before erosion wears them down) will be covered in snow and ice, and once again will reflect sunlight. The net effect is that the total amount of energy from the Sun available to the Earth is reduced a little, so that the globe as a whole becomes slightly cooler. This minor cooling, in itself, is unlikely to be sufficient to precipitate an ice age but, where other circumstances are favourable, may be just enough to tip the balance.

So we must look around for phenomena that occur every 150 million years or so. One suggestion has been that the energy output of the Sun itself fluctuates over a long period. This is a reasonable hypothesis. We know of other stars whose brightness changes periodically, over a timespan of hours, weeks or years. That stars can vary over much longer periods seems likely. To take a single indicative example that this might be so we can turn our attention to a star cluster named by the ancient Greeks the Pleiades, or Seven Sisters. One sister, Merope, was represented by a fainter star than the rest because, according to the myth, she took as mate a mortal, Sisyphus, rather than a god, like the others. The interesting point is that the Greeks were certainly able to see Merope with the naked eye, whereas today this is virtually impossible. Of course, it is feasible that the Greeks were more sharp-sighted than we are, and certainly their skies were clearer; but it is equally likely that Merope has dimmed a fair deal over the past few thousand years – in other words, that it is a variable star with a period of at least several thousand years. From here we can suggest that, perhaps, *all* stars vary their output periodically, but that for obvious practical reason we have no way of knowing that this is so in the case of stars whose periods are sufficiently long. (Who could set up a series of astronomical observations lasting one million years, let alone 150 million?)

The putative variability of the Sun's output is an intriguing idea; the hypothesis is to a slight extent supported by 'the mystery of the missing neutrinos'. The fact that the Sun is producing far less of these enigmatic particles than expected could well be a symptom that it is going through a quiescent period – although it could equally well be that there is something wrong with our theories concerning what makes the Sun shine.

A different suggestion is that the rate of the Earth's spin periodically slows down and then speeds up again, the 'slow' times coinciding with ice ages. However, it is not immediately obvious why this should be, and it is difficult to conjecture a mechanism whereby, once slowed, the spin could then be speeded up again.

Another factor that must be taken into account is that our Sun is not motionless in space; instead, it travels around the centre of the Galaxy once every 220 million years or so. In the process, it regularly encounters regions where space contains a more than usual amount of matter in the form of gas, dust and other debris. This has various effects. For example, the amount of the Sun's light reaching the Earth is reduced at such times, although it is hard to guess whether or not this would be significant. Perhaps more important, the Earth would be subjected to a greater than usual bombardment of material from space. This would make little difference except on the occasions when, by ill luck, one or more such items of infalling material happened to be big. As we have seen, the impact of a very large meteorite or cometary nucleus caused sufficient damage to our global climate that the dinosaurs and much of the rest of the existing lifeforms were extinguished. Experiencing several such impacts in a comparatively short period of time might well be sufficient to trigger an ice age – assuming the other circumstances noted above were suitable.

However, as will be evident, we are really still guessing as to the true cause of ice ages. One point favouring all theories that rely on changes affecting the Solar System as a whole, rather than changes confined merely to the Earth, is that the planet Mars is currently, like our own world, in the throes of an ice age. Various surface features on that world seem to show for certain that the planet has, in the past, enjoyed running water, but today there is none to be seen. In all likelihood, the water is frozen under the ground (there is also a fair amount in the planet's ice-caps). That both planets should be suffering ice ages simultaneously may, of course, be just a matter of coincidence; but it seems more likely that there is a common cause.

Will Linear A Ever Be Deciphered?

The forms of written languages undergo constant evolution – as an example, we need only compare a piece of medieval script with a modern printed newspaper. During its 1100 years of ascendance, from about 2500BC to 1400BC, the Minoan civilization of Crete used two forms of script. First, as with many civilizations, came hieroglyphics. Then, around 1700BC, the Minoans developed from the hieroglyphic script a new syllabic form, today called Linear A. When the Mycenaeans invaded Crete they adapted Linear A to produce a revised form, Linear B. Both scripts have been preserved on clay tablets, most of which are inventories; however, only a few hundred examples of Linear A exist, whereas there are some thousands of surviving tablets bearing Linear B.

Deciphering the scripts of the ancients is not an easy business, especially if those scripts do not form part of the 'family tree' of any currently existing script. For example, the meanings of ancient Egyptian hieroglyphs were unknown until, by chance, a 'dictionary' came to light. This was the famous Rosetta Stone, discovered in 1799. On it was inscribed the same text in three different scripts – Egyptian hieroglyphic and demotic as well as, most importantly, Greek. The British polymath Thomas Young (1773–1829), using the fact

LEFT *The Rosetta Stone.*
ABOVE *A statuette of the Minoan Snake Goddess discovered at Knossos.*
RIGHT *The queen's apartments at Knossos.*

that the name of Ptolemy appeared several times, was able to decipher part of the hieroglyphics. After Young's death his work was carried on by the French Egyptologist Jean-François Champollion (1790–1832). However, without the lucky discovery of the Rosetta Stone, it is likely that Egyptian hieroglyphics would still be a mystery to us.

No such 'dictionary' exists in the case of Linear A and Linear B. However, a brilliant young British scholar, Michael Ventris (1922–56), recognized in 1952 that Linear B was the written version of a language akin to Greek, and thanks to this inspiration made swift progress in deciphering it. Tragically, he died in a road accident just four years after starting. His work on Linear B was carried on with flair by John Chadwick (b. 1920). Ventris died before he had had the time to turn his attentions to Linear A, about which he is reported to have had some initial ideas.

Our inability to penetrate Linear A is something of a mystery. After all, as we've noted, Linear B is a development from the earlier script, and so the decipherment of one should provide a very useful clue to the other. Perhaps we shall be lucky and discover another 'dictionary' like the Rosetta Stone.

Why Did The Ancients Build Megalithic Monuments?

The use of very large stones – megaliths – either singly or in combination was widespread among the ancients, and it is clear that our ancestors must have regarded the practice as extremely important. After all, the transportation of such huge and heavy slabs of rock, often over long distances, must have required an incredible investment of effort on the part of any society possessing only rudimentary technology. The subsequent erection of the monuments – for example, capping the uprights of Stonehenge – likewise cannot have been easy. Indeed, some people have suggested that it was so difficult as to have been impossible, but this is to underestimate the intelligence and ingenuity of our forebears. Modern engineers have demonstrated, at least on paper, construction techniques that would certainly have been possible using the equipment available to Neolithic and later cultures. Which is far from saying that the task was *easy*.

That these monuments were vitally important to our ancestors is, therefore, obvious. However, the motives for

LEFT *The Sphinx with, behind it, the pyramid of Khafra, son of Cheops.* **ABOVE** *Some of the stones at Carnac, Brittany (France).*

building them are, in most cases, very much less so. We can understand why the Egyptian culture built the Pyramids: self-aggrandisement on the part of the pharaohs. After the first of them, the Step Pyramid, had been built (probably by Imhotep) for the pharaoh Zoser in the third millennium BC, all the other pharaohs wanted one too. Since the pharaohs were gods on Earth, the luckless culture had little choice but to kowtow to their whims, and vast amounts of expense and slave-power were devoted to the construction of these monuments. The rather dissimilar, and much later, pyramids of South America were probably largely religious in origin; if one wants to build a pretty impressive temple to one's gods, the pyramid is as good a shape as any to choose.

Elsewhere, in places as diverse as Western Europe and Easter Island, the motives are little understood – and still hotly debated. For a long time it was assumed by archaeologists that here, too, the purpose was religious. However, this was before the astronomers, notably Professor Alexander Thom, got in on the act. Thom examined countless stone circles and other megalithic monuments with an almost monomaniacal zeal, paying particular attention to the alignments of various of their features with events of astronomical importance – for example, the position on the horizon of the rising Sun at the winter solstice. It was perhaps not surprising that he found such alignments: the dating of events such as the solstices and equinoxes is very important to an agricultural society dependent on the cycle of the seasons. The importance is certainly practical; by extension it is likely to become religious as well.

But Thom did not stop there. He began to seek further alignments which might relate to far more subtle astronomical changes – and found them. For example, there was the matter of lunar *standstills*. Because of the way that the Moon moves in its orbit around the Earth, its place of rising on the horizon varies a little, each night moving a tiny amount further north until its position of rising seems to come to a halt (a standstill) before moving back southwards again. The whole cycle, from north to south to north, takes 18.61 years. However, things are not that simple. Because the period is

RIGHT *Stonehenge, England: sunrise at the midsummer solstice.*

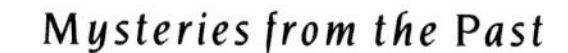

not a whole number of years, the next standstill will occur at a different season than its predecessor. Here lies the rub. If the standstill occurs in midwinter, the rising point will be at its most northerly; if in midsummer, the point is at its *most southerly* – and the picture is even further complicated during the intermediate times of year! The effect is, therefore, an infernally hard one to spot ... and yet Thom found more alignments with standstills than could realistically be explained by chance.

Many archaeologists were unable to swallow Thom's arguments – some because they were ignorant of astronomy and others, perhaps more reasonably, because the notion that our 'primitive' forebears could have such an advanced observational astronomy was anathema to them. When, around 1963, Gerald Hawkins applied a computer to the various alignments of features present at Stonehenge, finding their astronomical connotations, the storm burst. Thom could safely be publicly discounted as an eccentric (he was indeed something of an eccentric in many of his personal habits) and a generally irrelevant gadfly, but it was rather more difficult to depict Hawkins's computer in the same light. Entrenched archaeologists, desperate to refute the 'heresy' by any means, came out with some astonishing counter-arguments. Perhaps the most ludicrous was this:

- Hawkins required a computer to 'decode' Stonehenge, showing its function as an astronomical observatory
- our forebears did not have computers
- therefore, how could they have constructed Stonehenge as an astronomical observatory?

Bumble-bees do not have advanced knowledge of aeronautical science, yet some extremely complicated maths is required to explain why it is that a bumble-bee can fly. Does this mean that bumble-bees cannot fly? Obviously not. Yet this was the sort of argument used by various archaeologists, notably Glyn Daniel, who refused to accept the findings of Thom, Hawkins and a growing number of others.

In more recent years the controversy has died down a little, although it has still not ceased. It seems probable that, to our ancestors, there was no clear boundary between fields of endeavour that we now consider to be quite separate – for example, astronomy and religion. The significance of the

Sun's activities was extreme, because they affected the food-supply. It is a small jump from this to the veneration of the Sun as an important god. The behaviour of the Sun was therefore well worth watching, because it could influence the coming harvest. It would therefore seem to make a lot of sense to construct edifices that would allow you to observe the Sun's behaviour because, as a god, he might angrily starve you or beneficently reward you with plenty. The megalithic monuments can therefore be regarded in part as religious structures, in part as astronomical observatories, and in part as clocks, allowing our ancestors to record precisely the various cyclical changes of the celestial bodies.

These are speculations: to call them any more than that would be foolhardy. Nevertheless, we can conclude that many of the megalithic monuments had some sort of astronomical import (and this includes some aspects of the Egyptian Pyramids); as to their religious meaning we can only guess. A further mystery is this: if we assume that the motive for building the monuments was a mixture of religion and astronomy, which way round did the process operate? In other words, did the religious belief stimulate the precise astronomical observation, or was it that the study of the skies gave rise to beliefs about the gods?

This question is impossible to answer. If we look at, for example, Graeco-Roman mythology, we discover that the constellations in the night sky can be directly equated with gods, goddesses, heroes and other mythological notables. Did the ancients watch the heavens and, from what they saw there, construct their whole complicated pantheon? Or was it that they developed their legends of the deities and then 'saw' those deities portrayed in the sky? Why did the ancients identify the planets with such important deities – Mars, Venus, Jupiter and the rest? For that matter, why was the Moon – by far the most obvious object in the night sky – equated with only a minor deity, Diana? Would the whole story not have made more sense had Jupiter been identified with the Sun, Venus with the Moon, and so on?

The reasons for this apparent perversity are probably now lost to us for all time. However, who knows, perhaps someday we'll come across some tablet or document that will make the whole thing clear. Until then all we can do is wait . . . and wonder.

LEFT *Another shot of Stonehenge. Note the 'dagger' and 'axe' marks on the face of the upright in the centre.*

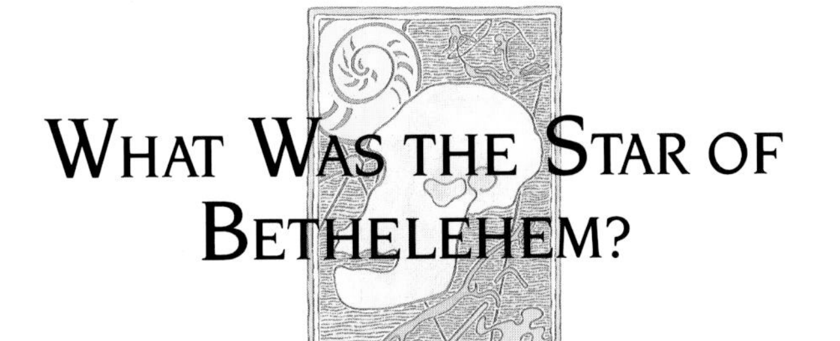

What Was the Star of Bethelehem?

Not so many years ago, this would have been a non-question: Christians believed that the Star was one of God's miracles, while scientists, by and large, believed that the whole story was a myth. More recently, these hidebound attitudes have changed, in that educated, intelligent Christians accept that there is little need for God to perform a miracle (unless, moving in mysterious ways, He particularly wants to) when He can create the same effect using natural processes; while scientists have begun more and more to recognize that there may be more truth in the mystical writings of the Christian and other religions than they have previously been prepared to admit. This latter point is not intended to imply that scientists have suddenly 'got God'; rather that they have generally refused to admit that, if some event is widely reported in religious writings, there is a very good chance that it did in fact happen.

Leaving aside the straightforwardly miraculous, there are two major possible explanations for the Star of Bethlehem. One is that a comparatively nearby star in our Galaxy exploded as a supernova, shining millions of times more brightly than usual. A second is that a rare conjunction of certain bright planets created the image of a very bright star. (A third possible explanation is that the whole story was made up, to impress unbelievers, by the apostle Matthew or whoever wrote the gospel ascribed to him. This hypothesis does not seem to have a lot going for it because, at the time Matthew – or whoever – was writing, there would still have been a few people around who could say something like 'I don't remember any star!')

The idea that the Star might have been a supernova was touchingly treated in a 1955 story by the science-fiction writer Arthur C Clarke, 'The Star'. He was not, so far as we can tell, putting forward the idea as a serious theory: his purpose was to tell a good story. Nevertheless, the idea is extremely plausible. We can look at it from both agnostic and religious viewpoints. The appearance in the skies of a bright 'new' star would certainly have influenced the ideas of our astrology-prone ancestors: this was a sign of major import. The coincidence of the Star's appearance with the birth of a self-proclaimed Messiah such as Christ would, certainly, have increased his 'street cred': as one of a succession of putative Messiahs and/or revolutionary leaders (all, so far, failed), he could claim that his birth had been heralded by this unusual celestial event. If we decide to be less cynical, we could suggest that God triggered a supernova at exactly the right time to draw the Magi towards the place where His son was being born.

It is difficult to establish which of the two scientific theories could hold sway. In recent years the idea that the Star was a supernova has been propounded by, among others, AT Lawton, a past President of the British Interplanetary Society. He notes that, although nearby supernovae are somewhat rare, an event such as this occurring in 4BC would make a lot of statistical sense. Of course, he has very little proof (how could he have?), but his various arguments are very persuasive.

Equally so are those of David Hughes, another distinguished astronomer and the author of *The Star of Bethlehem Mystery* (1979). Hughes points to a conjunction (appearance in the same point of the sky) of the two major planets Jupiter and Saturn in 7BC, a reasonable year for Christ's birth. The brilliance of this conjunction would certainly have been important to ancient astrologers: the three Magi, on the reasonable assumption that they were students of the skies, might very likely have followed the 'star' to its 'source'.

The ideas of Hughes are immediately provable: there was indeed a conjunction of the two great planets in the year he identifies. But there is a problem: the astrologers of the ancient world were certainly *au fait* with the behaviour of the planets and so, while they might have been excited by the conjunction of Jupiter and Saturn, they would not necessarily have regarded it as too desperately important. A supernova, by contrast, would have been a 'one-off' phenomenon – and hence much more likely to be the herald of the arrival of someone of importance.

Discovering the remnants of the supernovae of the past is a chancy business. When we are lucky, the orientation of spin of the resulting pulsars is such that we can place and date the initial explosion with some accuracy. Most of the time, though, we are unlucky: the supernova remnant is invisible to us, although we may be fortunate enough to observe the expanding cloud of gases and other matter receding from the pulsar. So far, no one has found a possible candidate for the supernova that might have represented the Star of Bethlehem.

LEFT *Detail of a 16th-century woodcut, possibly by Holbein, showing an astronomer observing a celestial body. Early astronomers did not have telescopes to assist their observations, but nevertheless they had a full knowledge of the skies.* **RIGHT** *A Portuguese sunset. One little discussed possibility is that the Star of Bethlehem was in fact simply the setting Sun, which the Magi followed westwards.*

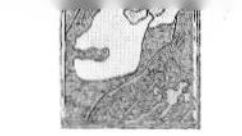

Kr
A
Hg

PART TWO

MYSTERIES OF LIFE

Life itself is something of a mystery. Most chemists will tell you that, once chemical reactions become complex enough, the end result will be life. But this explanation is not sufficient to explain the phenomenon: there is a definite difference between, shall we say, an amoeba and a crystal, yet the crystal grows and (in a very limited sense) reproduces – two activities that are generally assumed to be characteristic of living material.

So what is it that differentiates life from the nonliving? The distinction is not clear: at the same time that we recognize it, we cannot define it. We have similar difficulties when we look at such aspects as soul and intelligence: these have a profound effect on our daily existence, yet we have problems whenever we try to quantify them. After all, what is intelligence? It seems to us to be an evolutionary survival factor: the more intelligent a species, the more likely it is to survive in the long term; this is a reasonable hypothesis, but it can hardly be considered as any more than that.

When we look at the mysteries of life we must look also at phenomena that seem, on the face of it, to have little to do with the way that life works.

LEFT *Artificially grown crystals of one of the alums (aluminium ammonium sulphate).* **BELOW** *The great 19th-century scientist Friedrich Wöhler who prepared the first organic compound derived from inorganic chemicals. This was the death-knell for theories that organic chemicals possessed some 'vital force' unavailable to the inorganic chemicals.*

Do Human Beings Have a Soul?

Aristotle, who lived during the 4th century BC, said that there was a fundamental 'living principle' – or 'life force' – that distinguished living from nonliving material. The gods breathed 'vitalism' into living creatures, thereby giving them the quality of life. This life-principle was very important to the early alchemists: they saw it as so real that, not only did they consider all entities to be made up of differing proportions of dead matter and life force (spirit), they tried to use the spirit as, in essence, just one more chemical. As far as the alchemists were concerned the spirit of human beings and other living creatures was a quantifiable characteristic. Living creatures had 'vitalism': this was the distinction between them and the nonliving. Even the important German chemist Georg Ernst Stahl (1660–1734) supported these ideas.

In 1828 the theory started to fall to pieces. In that year the German chemist Friedrich Wöhler (1800–92) found a way of producing from inorganic materials the chemical urea; always beforehand this had been assumed to be a substance that could be produced only by bodily (biochemical) reactions. Later, in the 1840s, Emil Du Bois-Reymond (1818–96) showed experimentally that the impulses travelling along nerves are rather like the flow of electrical currents along a wire (the similarity is in fact even closer than he imagined). Later, in 1894, Max Rubner (1854–1932) demonstrated that the quantity of energy extracted from food by the body is largely determined by the laws of thermodynamics. The 1896 discovery by Eduard Buchner (1860–1917) that fermentation could carry on in the absence of living cells seemed like the final nail in the coffin.

But the lid of the coffin refuses to stay nailed down. One very important reason concerns Kirlian photography. Such photographs (named for Semyon and Valentina Kirlian, who explored the technique) seem to show auras – glowing

RIGHT *Georg Ernst Stahl, a chief promoter of the phlogiston theory; he was also a supporter of vitalism and an exceptionally sloppy scientist.*

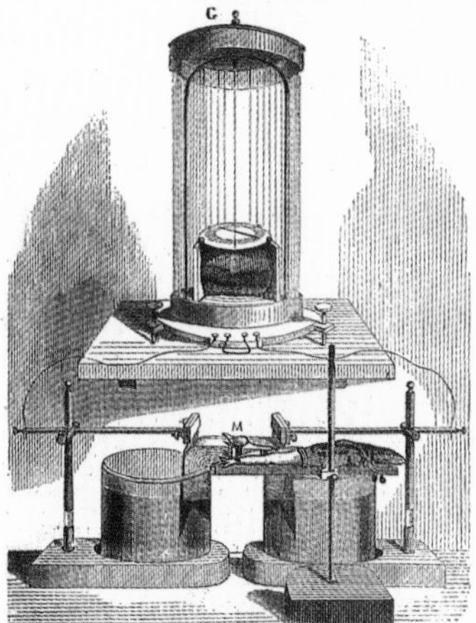

ABOVE *Emil Du Bois-Reymond, the 19th-century German physiologist who showed that muscular nerve impulses were akin to electricity.* **LEFT** *His equipment set up to demonstrate this fact using an unfortunate frog.* **BOTTOM** *Du Bois-Reymond demonstrating his apparatus for investigating muscular current and electric potential.*

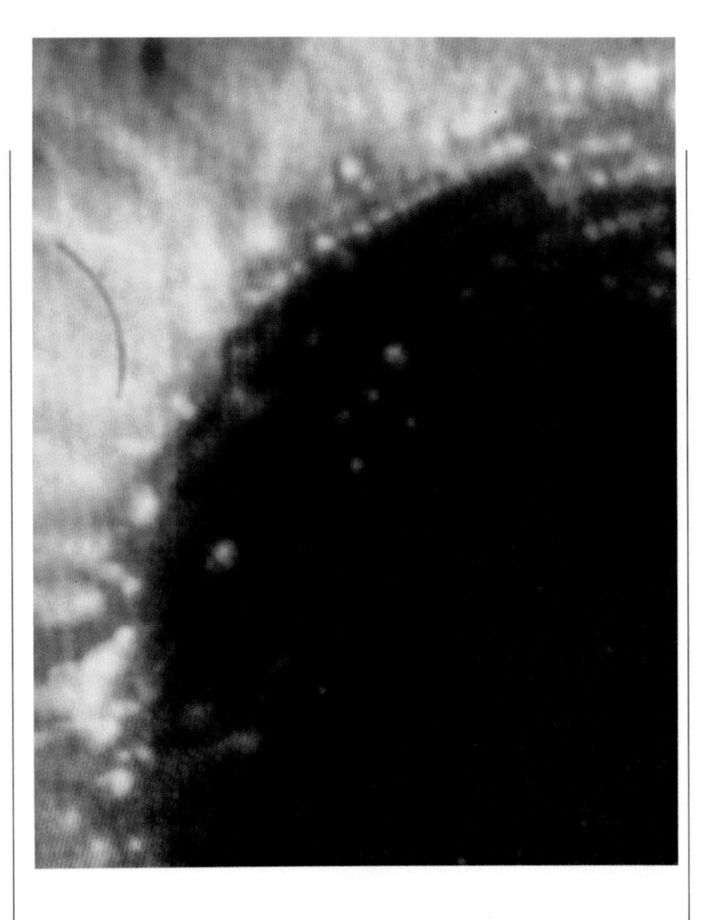

ABOVE *Part of a Kirlian photograph of the fingertip of a drug addict. There could hardly be a stronger contrast between this and the picture* **RIGHT** *which shows a Kirlian photograph of a healthy person's fingertip.*

haloes – around the margins of living and even recently dead organisms. Could these be photographs of the vitalistic ('soul') field of living creatures? Might it not be that the glowing auras are in fact photographs of the soul?

We have to reserve judgement. It seems not unlikely that human beings – and other living organisms – are surrounded by electromagnetic fields that could be picked up by cameras, assuming the circumstances were favourable. To leap from this standpoint to the assumption that we have an everlasting soul is to leap a very long way – probably too far.

Nevertheless, we can perceive that there is a difference between living (or recently dead) and nonliving material, and the Kirlian field is in some way involved with this. Such a differentiation could be as fundamental as life's ability to reproduce. Of course, to use such an argument to support vitalism would be specious: it is a nonsense to believe that only the presence of a 'vital principle' can confer life.

As for the soul? All we can do is guess.

Are There Other Intelligences?

Human beings have, traditionally, had a very high opinion of themselves. For millennia it was accepted that the Earth was at the centre of the Universe, despite all the contrary evidence, on the grounds that no other site would be fitting for the world on which the 'chosen' (whatever their religion) dwelt. Even today, fundamentalists envisage a Heaven barred to animals because, after all, only human beings were created in God's image and have souls. The idea that there might be intelligent lifeforms – indeed, civilizations more advanced than our own – on other worlds has been resisted vociferously by fundamentalists of various religions and, astonishingly, by a few scientists. The fact that even on our own planet we have creatures – the dolphins and whales – whose intelligence is at the very least comparable with our own makes little impact on people who deny the possible existence of intelligent life on other worlds.

One assumption is that any intelligent species will in due course develop a technology, just as we have done. This assumption is not borne out by any evidence – after all, the dolphins have no perceivable technology – but it is fair to guess that at least *some* intelligent species will develop an advanced technology. Those who rebut any ideas of extraterrestrial civilizations ask what seems at first sight to be a very salient question: 'Where *are* they all?' If the Universe is teeming with technological civilizations, why have they not come to visit us? At the very least, why have we not picked up traces of their domestic radio and television?

Why not, indeed? Yet, at the same time, why do we assume that a technological civilization will have the slightest desire to quest among the stars? We generally think that our own civilization does, but this belief is barely matched by governmental investment in the overall effort. Anyway, interstellar travel would be amazingly expensive – perhaps too expensive for most civilizations to contemplate. And why should we expect to have detected leakages from the domestic radio and television transmissions of our hypothetical aliens when we have made no effort to do so? If programmes aiming to pick up directly beamed 'contact' messages from other worlds are starved of funding, what possibility is there of anyone picking up the incredibly faint traces of extraterrestrial radio leakage? A further point concerns time. Despite the fact that our genus has been around for some four million years, it is only in the past century or so that we have discovered the knack of using radio signals. It is surely rather a lot to ask an alien species, perhaps 50 light-years away, to pick up the faintest of traces and then instantly respond. Even if the species did so, we would have little or no chance of picking up the message, because – to repeat – we are not looking out for it.

All discussions of extraterrestrial civilizations therefore have to be purely speculative. Some of the speculations have been extremely silly – and this is not to talk solely of the outpourings of the more sensationalist flying-saucer fans and their 'little green men'. For example, at a National Academy of Sciences meeting in Green Bank, West Virginia, in 1961, the otherwise distinguished US radio astronomer Frank Drake produced an impressive mathematical formula which purported to allow us to calculate roughly how many civilizations in our Galaxy would have developed a technology sufficient to allow them to make radio communication with us. When the equation is worked through, we discover that there should be at most 50,000 and at least 40 such civilizations. This is very impressive, until we start looking at some of the factors in the equation:

- the average rate of formation of stars in the Galaxy
- the fraction of stars that have planets
- the number of planets each star might have where the environment is suitable for life
- the fraction of such planets where life actually does appear
- the fraction of *those* planets where life leads to the development of intelligence
- the fraction of civilizations that develop the ability and desire to communicate with others
- the average lifetime of such civilizations

Reading down this list, we find that we have less and less possibility of being able to produce adequate figures to slot into the equation. We have a rough idea (although we may be mistaken) of the rate at which new stars form; we suspect that many, if not most, of them have planets – and so on until we come to the question about how long our 'ideal' technological civilization might expect to endure, to which any answer must be the most unfounded of guesses. Furthermore, is it not overwhelmingly likely that there are other factors involved of which we know nothing, or which we simply overlook? It is hard to find any reason to ascribe the slightest scientific validity to the Green Bank formula.

The problem with any such guesstimate is that, of course, we are basing our calculations on a statistical sample

LEFT *A metallic meteorite discovered in Mexico. Such meteorites give us few clues as to the prevalence of life on other worlds; however, members of a different class of meteorites – the carbonaceous chondrites – contain organic chemicals, leading us to believe that the building blocks of life must be fairly commonplace.*

of one. If a newspaper commissioned a political poll based on the opinion of a single person it would immediately become a laughing-stock. Scientists attempting to calculate the number of extraterrestrial civilizations are in exactly the same position: the only evidence we have for life anywhere in the Universe is confined to this planet.

Or is it? Some meteorites betray traces of proteinoid globules, the precursors of living cells. There is a great deal of suspicion that these may simply be the products of biological contamination after the meteorites have landed; but, if this is not the case, it would seem that life is likely to be rife throughout the Universe. And that, of course, leads us back to the question: 'Where *are* they all?'

The question acquires a new relevance when we look at the propositions of some of the CETI theorists (CETI = Communication with ExtraTerrestrial Intelligence). The US scientist Ronald Bracewell seems to have been the first to have suggested that sending or listening for radio signals is a haphazard business and almost certainly doomed to failure, because the task of covering the millions of 'likely' stars is well-nigh impossible. He proposed that, instead, we should pepper the skies with computer-controlled probes. This would be an expensive and very long-term project, but sooner or later one of the probes would encounter an extraterrestrial civilization – assuming there are any – and communications could be set up.

The idea has been expanded, notably by the Scottish CETI theorist Chris Boyce. Sending out a sufficient number of Bracewell-style probes would be prohibitively expensive. However, if such probes were programmed so that, each time they encountered a planetary system, they located a suitable asteroid and from its raw materials built several replicas of themselves, which were then despatched to further stars, the chances of success would be staggeringly increased. Such von Neumann probes, as they have been dubbed – after the US mathematician John von Neumann (1903–57) – would obviously have to possess a more than rudimentary machine intelligence. The likelihood of their establishing contact with an extraterrestrial civilization would therefore be yet further increased, because probes emanating from different civilizations would, with luck, run across each other, exchange information, and in due course establish communications between their parent cultures.

This approach has been described by its supporters as the 'uniquely logical' method of attempting to establish CETI. Some scientists who think that there is little likelihood of there being any other technological civilization in the Galaxy agree that the approach is indeed 'uniquely logical' – and have gone on from there to support their own case. After all, they point out, if there were other civilizations, the asteroid belt of our own Solar System should be packed out with von Neumann probes. To this there is an instant response: it may be, but we don't know because we haven't looked. (Also, of course, if every civilization adopts that attitude, then none of them will send out von Neumann probes . . .)

If we cannot make any sensible estimates as to the existence of intelligence on other worlds, at least we can look at non-human intelligences back here on Earth. There are two categories to be considered: animals and machines.

When we assess animal intelligence we tend to take a very anthropocentric view; that is, our yardstick of a creature's intelligence is how closely it resembles our own. This is not a

LEFT *A fixed radio interferometer aerial, 440m long, at the Mullard Radio Astronomy Observatory, Cambridge, England.*
BELOW RIGHT *A group of dolphins. Some scientists believe that these marine mammals may be more intelligent than human beings.*

Even when intelligence is measured in human terms, some animals show up rather well. A number of dolphins have been taught to communicate in rudimentary English. Their vocabularies are small, but even so the achievement is almost miraculous: how many human beings have learned to speak and understand dolphin-language? Chimps, whose vocal chords are not well adapted to human speech, have discovered how to communicate using deaf-and-dumb language; again, this is an incredible achievement. In both cases the limitations of vocabulary should not be seen necessarily to imply any lack of ability for abstract thought: these animals are using an alien tool to communicate with a species (us) whose intellectual make-up is quite alien to them, so it is hardly surprising that there are comparatively few areas of common ground that can be described using human words.

We can understand this more clearly if we imagine that the tables have been turned, and that we are trying to get by in basic dolphin-language. Dolphins have a sense which we do not have: using something akin to our sonar, they can 'see' their surroundings holographically (in true 3D, rather than the effect of 3D our stereoscopic vision creates). There are no words in any human language that can adequately describe the experience of 'seeing' something in this way –

useful approach, because it assumes that all forms of intelligence are of the same qualitative type. Any attempt to *quantify* the relative intelligences of species – and even of human beings from different cultural backgrounds – is quite patently misguided: one is trying to apply the same measure to two qualitatively different things. Even within the same culture, comparative tests of intelligence are on shaky ground; as has often been observed, all that IQ tests are good at measuring and demonstrating is the ability of individuals to pass IQ tests.

The point is not that we in any way deliberately do down the intelligence of animals – although we do this as well – but that it is hard for us to imagine the workings of forms of intelligence that have evolved to cope with environmental circumstances different from our own. The imperatives directing the evolution of intelligence differ between one species and the next. What is valuable to us, as a land-based animal, need not be valuable to a marine creature like a dolphin, and vice versa.

precisely because human beings do not have a holographic sense and so have never required the words to describe it nor the intellectual ability to be able to imagine it. (In exactly the same way, congenitally blind people can understand the physics of vision, but are unable to conceive what it would be like to experience seeing, for example, different colours. Helen Keller wrote very interestingly of her own conception of the different colours people described to her; the descriptions conjured up tactile sensations. She knew of the greenness of moss as 'velvety', the whiteness of lilies as 'soft'.)

How, then, can we evaluate animal intelligence? The answer is that we cannot, except in the vaguest of terms. There is little doubt that the average human being is more intelligent than the average cow – no contest! When it comes to comparisons with more overtly intelligent beasts – dolphins, whales, squids, chimps, gorillas, even rats – we cannot be so confident. None of these animals have the sort of intelligence that would enable them to function as successful human beings (ignoring the physical differences), but then we would not seem to have the kind of intelligence that would enable us to operate successfully in *their* environments.

Machine intelligence presents a slightly different case, because it is to a great extent governed by human intelligence: clearly, when we design an artificial brain, we create it in our own (intellectual) image. We expect it to be good at mathematics, particularly, and also at various other obvious human capabilities. However, we do not expect our computers to be capable of experiencing emotions, even though emotions are clearly a fundamental part of human intelligence. It is also a cliché that computers are incapable of creative thought – in other words, imagination.

BELOW *The computer centre at* CERN. **RIGHT** *The face of the future? A robot seals a seam at a car plant.*

This may change soon, though. Scientists talk in terms of 'generations' of computers, and the most advanced of the machines in operation at the moment are of the fourth generation. There is much speculation about the advent of fifth-generation machines which, it is proposed, will be capable of learning, extrapolation, ratiocination and imagination – and hence, possibly, of something that we would recognize as emotion. A science-fiction dream? Perhaps not. In the early 1980s a government-funded Japanese project was set up to develop a fifth-generation computer. The Japanese have a habit of scheduling their 'breakthroughs' in advance, which may seem silly but in practice generally works. The project predicted that their big breakthrough would come in 1990. Interestingly, the only doubts that Western scientists seemed to express about the viability of the project concerned this deadline.

Although the machine intelligences we create will be based on our own form of intelligence and an extension of it, we can nevertheless expect them to be alien in many ways. They will have senses beyond any that we can conceive. In a way, computers already do. For example, the simplest desktop computer functions through 'perceiving' shifting patterns of electrons, something which we cannot do because we are not aware of electrons in everyday life. Who knows what thoughts may pass through the mind of a machine capable of consciously perceiving such patterns, and *thinking* about them?

COULD 'LIFE AS WE DO NOT KNOW IT' EXIST?

All life on Earth is based on the element carbon. This is no coincidence: carbon atoms are almost uniquely capable of linking up to form, with atoms of other elements, molecules of virtually any length and complexity. We note the word 'almost' in that sentence.

All life on Earth depends on water. Again this is no coincidence, because life began either in the oceans or in water-affected environments.

Finally, almost all forms of life on Earth rely on the presence of oxygen. Chemical rections in our body involving oxygen give us the energy we need to live (some types of bacteria, anaerobic bacteria, curl up and die when exposed to oxygen, but they are in the minority). This is still as we might expect, because oxygen is a very reactive gas and was from very early days prolific in the Earth's atmosphere, albeit probably in the form of carbon dioxide.

We can thus describe any lifeform based on the carbon-water-oxygen triangle as 'life as we know it'; the expression can be taken to embrace also the anaerobic bacteria. But is this the only basis on which life can exist?

In theory, no. Some biochemists have built up quite elaborate blueprints for forms of life utterly different from our own. For example, the role of carbon can be performed, although not as well, by the common element silicon. Like carbon, silicon atoms can form chains, although they are not as stable. (There are other reasons why complicated molecules based on silicon are less likely to occur, but these are beyond our present scope.) Silicon-based life would have to be very different from 'life as we know it'; to take just a single example, whereas we exhale carbon dioxide, an oxygen-oriented silicon lifeform would perforce have to exhale silicon dioxide (silica) – a painful experience! Nonetheless, in certain conceivable planetary environments, some form of silicon-based life might be feasible.

Is water so very important to life? As we've noted, all forms of life on Earth depend on water for the simple reason that life's very origin on this planet involved water. But what about other planets, where water may be a scarce commodity? In this respect particular attention has been paid to the giant gas planets like Jupiter and Saturn. Such planets cannot really be thought of as having a surface, in the sense that the Earth has; rather, we should regard them as being 'all atmosphere'. Common among the gases in these atmospheres are methane and ammonia. Ammonia, a molecule formed by nitrogen and hydrogen, seems unpromising as an analogue of water, but methane, formed by carbon and hydrogen, would seem to hold out a lot of hope. Once again, some scientists have worked out intriguing biochemistries for lifeforms oriented towards methane rather than water.

In fact, if – 4.6 billion years ago – one had to place a bet on which planet in the Solar System would sustain life, one would have been best advised to put money on one of the gas giants. These planets are big enough to be regarded as, in essence, stars that didn't quite make it. They generate a fair amount of internal heat and they are also riven by electrical storms of a ferocity that is hard to imagine. Both circumstances favour the formation of complicated molecules. A plethora of organic molecules does not, of course, mean that life will inevitably emerge, but it does increase the probability.

As an extra incentive to put money on the gas giants as havens of life we can note that each of them has an atmospheric layer where there is plenty of warm but polluted water. The circumstances in such layers would favour the emergence of 'life as we know it'.

Finally, we have to consider whether or not oxygen is essential to the evolution of life. We have the example of the anaerobic bacteria to show us that life can exist – indeed, may *have* to exist – without the presence of oxygen, but that is a slightly different concern: it seems likely that these bacteria evolved in an oxygen-rich environment and then adapted to ecologies lacking in oxygen. Could life originate in environments where there is no oxygen? Again we have to give the answer that, yes, possibly it could. One gas which has been investigated as an analogue of oxygen is fluorine. This gas is much less common than oxygen and very much more reactive (in fact, it is the most reactive of all the elements), probably too reactive to serve as the basis for any conceivable lifeform. For both of these reasons, fluorine-oriented life seems to be nothing more than a theoretical fancy. On the other hand, there are vast numbers of planets in the Universe, so there is at least a possibility that, somewhere, such a lifeform exists.

In all of these discussions we have been talking about lifeforms as physical entities – creatures who, like ourselves, have bodies. However, there is another way of looking at the phenomenon of life, and that is in terms of consciousness. As we have seen, it is hard to draw the line between a conscious human being and a conscious machine; likewise, we have to accept that *any* entity possessed of consciousness is in reality a living organism – whatever its physical characteristics. This idea was explored by Fred Hoyle in his seminal science-fiction novel *The Black Cloud* (1957); Hoyle suggested that a cloud of interstellar gas could develop a consciousness. The scientific underpinning of such an idea is necessarily speculative, but that is far from saying that it is a nonsense: such clouds are at least as complex, in terms of interacting molecules, as human beings. There is no obvious reason why they *cannot* be intelligent, and therefore describable as lifeforms. Which would represent 'life as we really do not know it'.

BELOW *The Horsehead Nebula in the constellation Orion. The nebula is a cloud of cool gas and dust; we see it silhouetted against the light of hot interstellar material. Below the brilliant blue star on the left we see another gaseous nebula.*

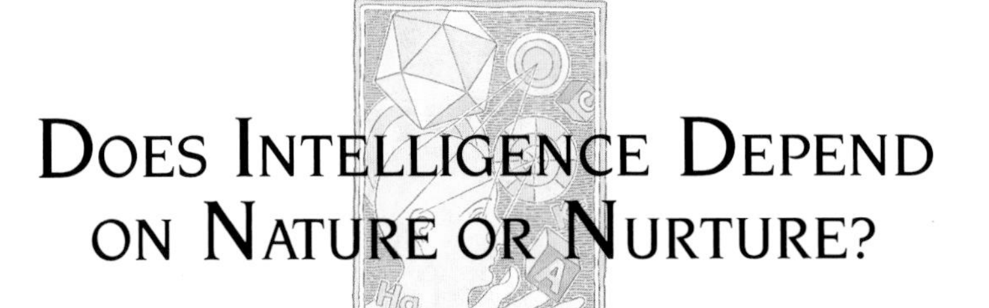

Does Intelligence Depend on Nature or Nurture?

Cyril Burt died in 1971. Throughout his lifetime he had been regarded as an excellently scientific psychologist who had shown that the level of a child's intelligence has little to do with the child's home environment; instead it is a product of the intelligences of the child's parents. Fairly good evidence has since appeared to show that Burt faked his results in order to bolster his case, although he still has some apologists.

The idea that a child's intelligence might depend solely on the intelligences of his or her parents seems commonsensical: all around us we see clever kids born of clever parents. However, the situation is not quite as clear-cut as it might seem. Clever parents are, naturally, going to give their children a more encouraging educational environment: to take a single example, they are not going to have the television set on at full volume while the unfortunate kids are attempting to do their school homework. It is now generally accepted that nurture – the environment in which a child grows up – is a more important factor than heredity when it comes to the intelligence of the resulting adult human being.

That nurture is more important than nature seems to be borne out by a study published in *Nature* in 1989. This showed that, on average, children adopted by high-earning

LEFT *The relationship between a child and both its parents, especially in the early years, seems to have a dominant effect on how well the child will be able to exploit her or his intelligence in later life.* **ABOVE RIGHT** *Television can be a powerful influence for education or, conversely, hinder it.* **FAR RIGHT** *The gift of reading is one of the most important a parent can give to a child.*

families had an IQ 12 points higher than similar children adopted by low-earning families – whatever the social class of the children's natural parents. Obviously it wasn't the money that made the difference: what was important was the fact that the wealth opened up all sorts of possibilities to the lucky children that simply were not available to adopted children of poorer parents.

The results of the survey are still controversial, so perhaps we should not allow them too much weight. Moreover, as Dr Matt McGue of the University of Minnesota commented at the time,

> ... WORKING-CLASS PARENTS CAN PROVIDE THEIR CHILDREN WITH INTELLECTUALLY STIMULATING EXPERIENCES AND PROFESSIONAL PARENTS CAN NEGLECT THE NEEDS OF THEIR CHILDREN ...

In other words, nurture is very important but heredity is probably significant, too. Or is that really true? We look at families like the Huxleys, the Darwins and the Russells and we begin to wonder.

One thought which throws a much-welcomed spanner in everyone's works is this: the parents of Einstein and Shakespeare were not noted for either their intelligence or their prosperity, yet their children seem to have done all right. In short, the whole debate about the relative contributions to a child's intelligence of upbringing and genetics is still wide open. It seems probable that a mixture of the two factors is involved, but we cannot be sure that this is the case and we most certainly cannot say which is more important than the other – let alone quantify (in terms of percentages) their relative importances.

We can, however, make a few generalizations. Book-oriented homes lead to book-oriented children: a child brought up in a home where reading is encouraged as a major activity is more likely to read voraciously from an early age. Such children may not be in fact more intelligent than their schoolmates, but certainly they are able to exploit more fully the intelligence which they do have, and possibly be better-informed. This is an advantage which they may continue to enjoy throughout life.

Is Immortality a Scientific Possibility?

The reasons for our mortality become obvious as soon as we stop thinking of ourselves as individuals. This is not a very palatable view: each of us likes to think that we are 'somebody special' – which is why each of us accepts intellectually that someday we must die but at a gut level believes that, in our own case, God, Fate or whatever will make an exception. However, if we look at the phenomenon of life in terms not of individuals but of the species as a whole, the necessity of personal death becomes obvious.

In order to survive, a species must be capable of adapting to changing circumstances. It can do so only through a constant turnover of individuals, whose sole purpose in being born is to reproduce and then die. Those products of reproduction – children – that are best adapted to the current circumstances will generally survive until they are of a suitable age to reproduce, passing on at least some of the characteristics favouring survival to their own children: those who are poorly adapted to the current circumstances will, in general, either die young or be unable to find a mate. Richard Dawkins has taken this idea further by suggesting that the sole purpose for human existence is to allow the further evolution of genes.

Evolution is generally regarded as a slow process, but in fact it works remarkably swiftly. To take a single example, it is reported that the grandchildren of the first European colonists of North America already had eyelid-shapes tending towards those of the Red Indians, which were better adapted to looking out over vast spaces in bright sunlight. Organisms simpler than human beings, like bacteria and viruses, mutate even more rapidly, because of both their simplicity and their rapid reproductive cycle.

Individual immortality would threaten the species as a whole – almost certainly with fatal results. If the immortals were capable of reproduction there would be a colossal population explosion: imagine what it would be like if every human being ever born were still alive! Alternatively, if the immortals were not capable of reproduction, their species would inevitably stagnate, be unable to adapt and in the end – through accident or otherwise – die out.

This is all very well, but most of us are interested in not general but *personal* immortality; others may come and go, but we ourselves, as individuals, would like to carry on living forever, please. Medical science – in particular, genetic engineering and spare-part surgery – is beginning to offer hopes of, if not immortality, then at least a major extension of lifespan. This may seem an attractive proposition, but in fact it is rather frightening – because it would be available only to some of us. We can conjecture immense social strife, as the 'haves' battle with the 'have-nots'. Similarly, it is reasonable to suppose that many of the haves will be among the most ruthless of society, and will pass on their inhumane ethic to their children, thereby perpetuating it: is this really the future we wish for our species?

A quite different way of extending humanity's average lifespan concerns genes and the age at which we reproduce. Most people are in their twenties or early thirties when they have children; the average age of first parenthood seems if anything to be falling. Any inbuilt coding – in the form of genes – that could affect us after our child-bearing period is irrelevant to the survival of the species: the genes involved might code us for death but, in terms of the species, this does not matter because we have already served our purpose by reproducing. Obviously the hypothetical 'death-genes' cannot come into action before later life. If, therefore, human beings as a whole progressively delayed the age at which they reproduced, the 'death-genes' would, over the course of the generations, be gradually filtered out – for the very simple reason that people most affected by 'death-genes' would die before reproducing, and hence would be unable to pass on the fatal genes to the next generation.

We must return to the fundamental question: is the extension of the human lifespan actually desirable? We already have a global population problem: do we really want to exacerbate it? Obviously not. Yet there is no reason – aside from short-sighted economics – why we should not increase our living area by creating new habitats for our species. These could be on other planets, in artificial space colonies or even beneath the Earth's own oceans. The possibilities are there: all we have to do is grasp them.

And, if we do, there are no external reasons why the average lifespan of all human beings should not be extended to many hundreds of years.

How Many Senses Are There?

The knee-jerk answer is that there are five; but, like most such answers, this one is wrong. In addition to the generally recognized senses of vision, hearing, touch, smell and taste, human beings are known to have an additional one, called proprioception or kinaesthesia, whereby we are aware of the relative position of our limbs, the tensions in our muscles and so on.

Clearly, psychics should talk about having 'a seventh sense' rather than the sixth! There has been a lot of debate, generally conducted outside scientific circles, about the possible existence of such a sense – or, to be more accurate, set of senses. We can term them collectively ESP (for Extra-Sensory Perception), thereby embracing the supposed abilities of telepathy, clairvoyance, precognition and so on. In so doing we are probably, assuming such abilities exist, mixing chalk with cheese: clairvoyance, for example, may have as little to do with precognition as vision has to do with touch. We do not yet have any way of knowing.

ABOVE *George Zirkle who, in the early 1930s, showed impressive scores in ESP experiments run by JB Rhine.*

LEFT *A 19th-century experiment intended to demonstrate the existence of telepathy. As we can see, there was very little 'science' involved in these experiments; they were treated rather like party tricks – which, almost certainly, they were.*

What scientific evidence have we for the existence of ESP? The sad answer is: very little, if any. Laboratory experiments carried out on the subject have proved, almost always, frustrating. Those that have seemed to indicate that ESP might be at work have generally later been discredited: either the subject of the experiment has been proven to be a charlatan or the experimental procedure itself has been shown to be flawed. For legal reasons it is impossible to mention examples of the former instance. As an example of the latter we can recall one case in which the famous 'scientific' investigator of the paranormal, JB Rhine (1895–1980), asked an experimental subject to try, clairvoyantly, to perceive the images on a set of cards that Rhine was turning up. At the time, Rhine was sitting in the front of his car while the subject was in the rear seat. However, Rhine was convinced that the astonishingly successful experiment showed a genuine result, because he had told the subject not to peek!

The scientific evidence may be lacking, but there is a colossal body of anecdotal evidence in favour of ESP – and there are few people who, in honesty, cannot recall one or more instances in which they themselves have had an experience which they can explain only in terms of it. The vast majority of such experiences can easily be interpreted in terms of orthodox science – usually statistics: most people, including myself, have only a hazy understanding of the workings of probability, and so perfectly commonplace coincidences seem to be outrageously improbable, and are therefore viewed as being in some way paranormal. Moreover, many widely reported accounts of astonishing experiences are elaborations or flat lies. Even so, some of the anecdotes cannot be so lightly dismissed. Out of the many examples available, here is one chosen more or less at random. It was recounted by Colin Wilson in *The Directory of Possibilities*:

JANE O'NEILL WITNESSED A SERIOUS ACCIDENT, AND WAS SO SHOCKED THAT SHE HAD TO TAKE SEVERAL WEEKS OFF WORK. AFTER THIS SHE HAD ODD FLASHES OF 'CLAIRVOYANCE'. ONE DAY, SHE VISITED THE CHURCH AT FOTHERINGHAY WITH A FRIEND AND WAS IMPRESSED BY A PICTURE BEHIND THE ALTAR. SHE MENTIONED IT LATER TO HER FRIEND, WHO SAID SHE HAD NOT SEEN THE PICTURE. LATER, TO SETTLE THE MATTER, THEY RETURNED TO THE CHURCH AND, TO O'NEILL'S SURPRISE, IT WAS A COMPLETELY DIFFERENT PLACE; IT WAS SMALLER, AND THE PICTURE WAS NOT THERE. SHE CORRESPONDED WITH AN EXPERT ON THE CHURCH, WHO TOLD HER THAT THE CHURCH SHE HAD 'SEEN' WAS THE CHURCH AS IT HAD BEEN 400 YEARS AGO, BEFORE IT HAD BEEN REBUILT IN 1553.

There is little reason to doubt O'Neill's honesty: clearly she had a very unusual experience. However, this does not necessarily mean that ESP was involved. As Wilson mentions, she had recently had a harrowing shock, so it is possible that she was hallucinating in some way and that, by chance, her hallucinations corresponded approximately with historical reality. Researchers into the paranormal would say, quite fairly, that such complicated explanations are actually more difficult to believe in than ESP. In this case, as in so many others, the truth is impossible to establish because the event was a personal one: the individual 'knows' what he or she experienced, but we can never, as it were, run through a tape-recording of the experience in order to try to establish what 'really' happened.

The existence or otherwise of ESP is one of the great mysteries of science, yet very few scientists show any interest

LEFT *Fotheringhay Church in Northamptonshire, England, where Jane O'Neill underwent what appeared to be a timeslip of four centuries.*

RIGHT *A telepathy experiment featuring a blindfolded Kuda Bux, who is reproducing the symbols drawn on the board by Lucie Kaye despite the fact that he cannot see them. Or can he? Most modern researchers into the paranormal set little store by such experiments.*

in it whatsoever. A major reason for this unconcern is that various fraudsters who have operated in the field have given ESP a bad name: underfunded scientists have far better things to investigate than a phenomenon which seems, on the basis of the available evidence, to be nothing more than a farrago. This is a pity, because the few scientists who have gone into the matter in rigorous detail have concluded that there at least *might* be something in it.

Certainly there is a lot of evidence that we have senses other than the 'official' six. Some people are always aware of the direction of north (as are migratory birds). A few people never need to wear a watch because they always know what the time is. Many of us find that we 'know' when we are being watched. Some people can 'feel' the presence nearby of massive objects; even if blindfolded they would run little risk of walking into a wall. An appreciable number of us have 'perfect pitch': we can identify a musical note exactly. All of us feel heat – we have a 'temperature sense'. The list could go on for a long time.

When we turn to the animal kingdom we discover a whole array of senses which we, as human beings, reckon we do *not* have. The use of sonar by bats, dolphins and others is a prime example: to describe it as merely an extension of the sense of hearing is simplistic, especially in the case of dolphins. Then there is the little understood sense possessed by certain fishes, whereby they can detect an object because of its minute electrical activity. Many organisms can 'see', at least fuzzily, using heat-sensitive cells – a handy ability for a night-time predator. This is another list that could go on for a long time.

In sum, there is no definitive answer to the question of how many senses there are: we have yet to discover all of our own, and even less do we know the full details of those enjoyed by some of the other members of the Animal Kingdom. There is even evidence that plants have senses, although we have no real conception of what those senses might be. The subject as a whole is a mystery for the reason which we touched on earlier in the context of ESP: we can never share another person's experiences. The same is true, but this time with a vengeance, when it comes to animals and plants. We can look at the way other creatures behave and from our observations infer the senses to which they are responding, but our inferences may be misleading us. For example, does an animal recoil from a naked flame because it can feel the heat or because it can 'see' the heat – or because it has some completely different sense that alerts it to the danger?

We do not know.

Do the Changes in the Developing Embryo and Foetus Reflect Our Evolutionary History?

As anyone with regular access to the colour supplements knows, the human embryo and foetus display marked changes during the nine months of gestation. An unborn child starts off as a tiny sphere, soon begins to look like a minute hamburger (complete with bun), and finally adopts the form of a large-headed, small-limbed human being.

The German naturalist Ernst Heinrich Haeckel (1834–1919) is generally regarded as the first to put forward the idea that these changes in the embryo retrace our evolutionary history. His catchphrase was that 'ontogeny recapitulates phylogeny', ontogeny being the sequence of events involved in the development of the individual embryo/foetus and phylogeny being the sequence of events involved in the evolution of a species.

Haeckel's hypothesis had no genuine theoretical underpinning – why *should* the changes in the developing embryo mirror the transitions involved in the evolution of the species? Nevertheless, it was important throughout the latter part of the last century and well into this one. This was because, despite its deficiencies in terms of theory, it seemed to be borne out so well by observation. We can look at a couple of points:

LEFT *A human foetus 8½ weeks after conception. Already the principal physical features of the future human being are clearly distinguishable.* **FAR LEFT** *At 12 weeks after conception the foetus is unmistakably a miniature human being.*

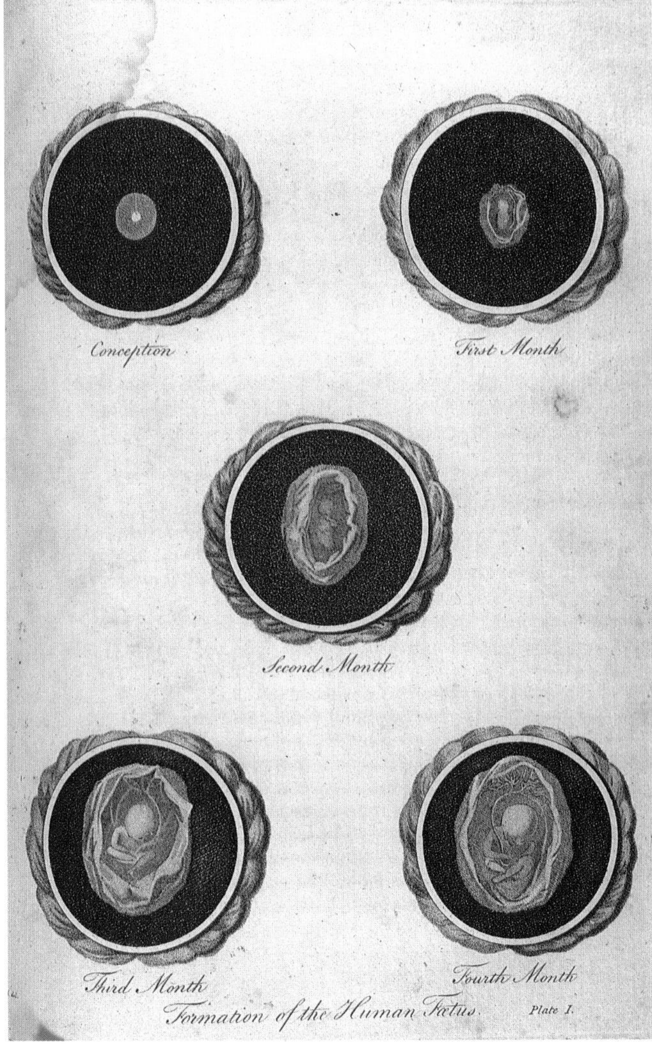

ABOVE *Ernst Heinrich Haeckel, the most serious proponent of the theory that ontogeny recapitulates phylogeny.* **ABOVE RIGHT** *A medical textbook's illustration from 1794 showing the development of the foetus during the first four months after conception.*

- in its earliest stages the human embryo has gills, as had our distant ancestors, which were marine creatures
- initially the embryo of any other mammal (such as a cat) is hard to distinguish from that of a human being, and only later do the necessary differences appear – this would seem to reflect the processes involved in divergent evolution

Haeckel's ideas are today generally discarded as a nonsense, but this is unfair and possibly misguided. Of course the parallel between ontogeny and phylogeny cannot be carried the whole way through – prehistoric human beings did not look like newborn babies – but there are various reasons why we might expect the early stages of embryonic development to have some relation to the early stages of our evolution. To take a single example, nature (otherwise often profligate) repeats any particularly useful trick it has learned over and over again. If nature has discovered the 'best way' of producing a sequence of changes that will result in an adult member of *Homo sapiens*, might it not recapitulate this sequence in the womb?

We cannot take such ideas too literally. Nevertheless, it does not seem ridiculous to consider parallels between the developing embryo/foetus and the evolution of our species. Whether these parallels are meaningful or merely coincidental is yet another mystery confronting science.

What Do Dreams Mean?

All of us dream. Some people believe that they do not dream; in fact they do, but they are unable to remember their dreams when they wake in the morning. Researchers have discovered that periods of dreaming during sleep are related to times of rapid-eye movement (REM) during which the eyes flick from side to side under the eyelids. You can see this happening if you watch someone asleep.

From earliest times it has been assumed that at least some dreams are important – bearing messages either from the gods or from the dreamer's subconscious, depending upon one's cultural context. These messages, it is widely believed, can give information about the future, the remote past or things taking place in distant parts of the world; at a more down-to-earth level, psychologists and especially psychoanalysts think that analysis of dreams can reveal details of a person's mental state, and have concocted numerous systems relating objects or events seen in dreams to aspects of the human psyche. Sex is the favourite topic: among the many symbols of sex are serpents, swords, running up a flight of stairs, swimming, flying . . . In fact, it is quite hard to think of a dream-image which cannot in some way be ingeniously linked to an aspect of the dreamer's sex-life, lack of it, or attitudes towards it.

All of us who recall our dreams recognize that they come in two different categories. These can be described as 'only' dreams and 'different' dreams. An 'only' dream has little effect on us; we may remember it in the morning because it was funny or for some other reason, but it does not disturb us in any way; most often, we recollect it only in the moments after waking and have forgotten it by the time we get out of bed. (An interesting way of countering this effect is to keep

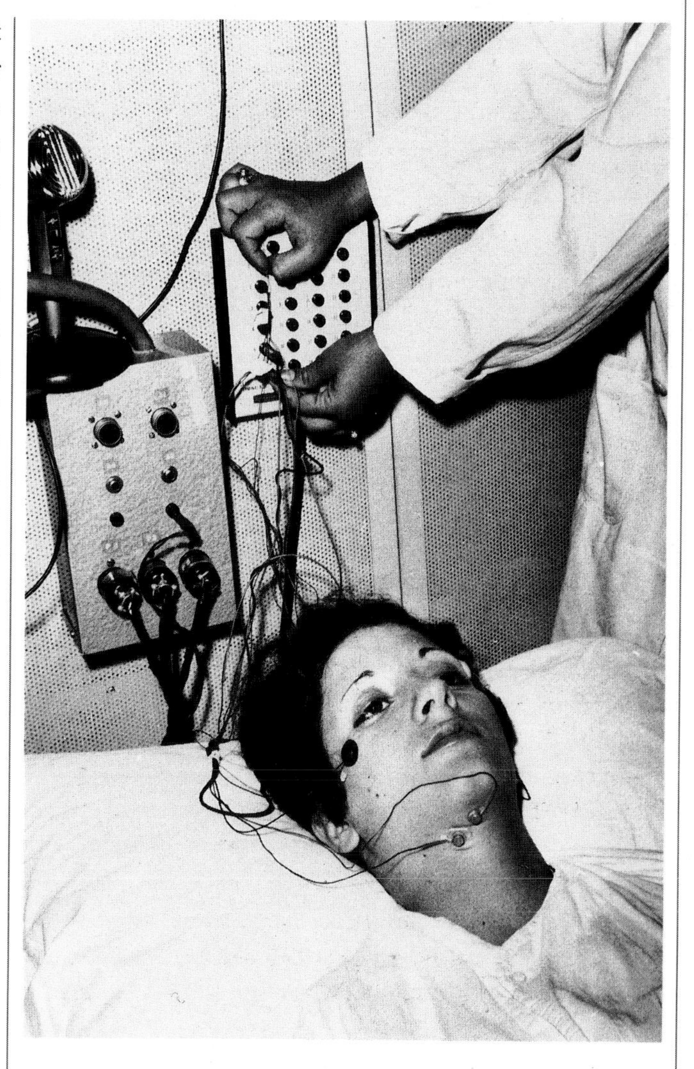

ABOVE *A dream-telepathy experiment being set up at Maimonides laboratory in Brooklyn, New York.*

LEFT *Robert Louis Stevenson and* **ABOVE** *Edgar Allan Poe, two writers who exploited their dreams to create works of imaginative fiction.*

by the bedside a 'dream diary' in which, as soon as you wake, you can jot down any memories you may have of your dreams. You will be surprised by how much more often you dream than you think you do.) A 'different' dream is a totally different beast from an 'only' dream. The prime example of a 'different' dream is a nightmare. Few of us are lucky enough to have escaped the shattering sensation of pure terror generated by nightmares: the experience can haunt our every action for days afterwards. But 'different' dreams need not necessarily be unpleasant; they affect our waking consciousness, certainly, but the effect may be one of pleasure rather than torment – as with a sexual dream. When people try to investigate the possible meanings of dreams, they focus on the 'different' dreams, regarding the 'only' dreams as just so much mental detritus.

It is possible that *all* dreams are merely mental detritus. This is not to say that dreams are unimportant. Far from it: a 'different' dream can profoundly influence the dreamer. What is being suggested is that dreams *in themselves* have no absolute meaning or import – so that the idea of analysing them in terms of standard sets of criteria becomes a nonsense – but that the reaction of the dreamer to his or her dream can be very significant. Imagine, for example, that two people have an identical dream – for the sake of argument we can conjecture that it concerns standing on an ants' nest. To one person this is an amusing fantasy, worth mentioning at the breakfast table, perhaps, but little more than that; to the other it is a horrific nightmare which lingers in the memory for days. The dream itself has no meaning; the effect it has on the second person tells a great deal about that person.

Nobody is yet sure what dreams actually are. We can describe them in terms of electrical activity in the brain, but this does not really tell us much. We all know, as dreamers, that dreams are genuine experiences, even if they cannot be recorded with a camera or a tape-recorder. Sometimes our dreams tell coherent stories, but more often they are a jumble that seems logical at the time but is revealed by our conscious memory as a nonsense. Some people experience 'lucid' dreams: they are aware, while dreaming, that that is exactly what they are doing, and in some cases can direct the course of the dream. Others dream when not fully asleep, either while waking (hypnopompic dreams) or while falling to sleep (hypnagogic dreams), and a few can do so lucidly – Edgar Allan Poe and Robert Louis Stevenson are two examples of writers who have gained inspiration through the use of lucid hypnagogic dreams.

We can look to one of Stevenson's non-lucid, non-hypnagogic dreams for a fine example of the way that day-time preoccupations can affect the dreaming mind, and vice versa. In his *Across the Plains* (1892) he wrote:

> I HAD LONG BEEN TRYING . . . TO FIND A BODY, A VEHICLE, FOR THAT STRONG SENSE OF MAN'S DOUBLE-BEING WHICH MUST AT TIMES COME IN UPON AND OVERWHELM THE MIND OF EVERY THINKING CREATURE. I HAD EVEN WRITTEN ONE, *THE TRAVELLING COMPANION*, WHICH WAS RETURNED BY AN EDITOR ON THE PLEA THAT IT WAS A WORK OF GENIUS AND INDECENT, AND WHICH I BURNED THE OTHER DAY ON THE GROUND THAT IT WAS NOT A WORK OF GENIUS, AND THAT *JEKYLL* HAD SUPPLANTED IT. THEN CAME ONE OF THOSE FINANCIAL FLUCTUATIONS . . . FOR TWO DAYS I WENT ABOUT RACKING MY BRAINS FOR A PLOT OF ANY SORT; AND ON THE SECOND NIGHT I DREAMED THE SCENE AT THE WINDOW, AND A SCENE AFTERWARD SPLIT IN TWO, IN WHICH HYDE, PURSUED FOR SOME CRIME, TOOK THE POWDER AND UNDERWENT THE CHANGE IN THE PRESENCE OF HIS PURSUERS. ALL THE REST WAS MADE AWAKE, AND CONSCIOUSLY . . . ALL THAT WAS GIVEN ME WAS THE MATTER OF THREE SCENES, AND THE CENTRAL IDEA OF A VOLUNTARY CHANGE BECOMING INVOLUNTARY.

Here we can see the two-way exchange between waking consciousness and dream-consciousness. Stevenson was deliberately seeking a plot that would allow him to explore an aspect of human psychology. It would seem reasonable to suggest that it was this waking preoccupation of his that caused his subconscious to produce the relevant dream. However, we can turn this line of reasoning on its head. Had he had the same nightmare at any other time, it might have had no particular meaning to him at all. *Any* 'different' dream he had around that time might have served to give him the 'vehicle' he was seeking. The 'meaning' of the dream seems therefore not to have been implicit in the dream itself but to have come from the interaction between its imagery and Stevenson's consciously controlled imagination – 'All the rest was made awake . . .'

There are various current theories which attempt to explain the phenomenon of dreaming. None are particularly satisfactory and all seem to be unprovable. They are worth mentioning here only because they are interesting. One idea is that we pick up all sorts of information during our waking hours without being consciously aware that we are doing so; the function of dreaming is to allow our brain to process all this information at the unconscious level. Another, closely related, hypothesis is that at the end of each day we have in our unconscious a sort of ragbag of bits and pieces of experience which our conscious mind has not had the time, opportunity or inclination to process; once again, the function of the dream is to deal with this material. A different notion concerns wish-fulfilment: in our dreams we can do with impunity things we would like to do in real life but cannot – make love with a Hollywood sex symbol or murder our boss. This is a very appealing theory but scarcely seems to be borne out by the facts; after all, we dream about all sorts of things which in no way do we wish to experience in real life. A very prosaic theory proposes that the electrical activity of the brain as we sleep produces the mental equivalent of white noise and that, just as we can make ourselves hear music in white noise, our unconscious can pick out a coherent story from the baffling array of visual images presented to it. Less scientific explanations of dreaming include messages from the gods, the ability of the spirit/soul to travel outside the body during sleep, and so on.

Dreaming does not seem to be confined to our own species: research on the subject would appear to indicate that animals likewise have dreams: we have all seen a sleeping dog twitch as it 'chases rabbits'. Charles Darwin, in *The Descent of Man* (1871), wrote about the matter:

ABOVE *One of William Hole's illustrations for Stevenson's* Dr Jekyll and Mr Hyde, *probably the most famous of all dream fictions.*

AS DOGS, CATS, HORSES, AND PROBABLY ALL THE HIGHER ANIMALS, EVEN BIRDS, HAVE VIVID DREAMS, AND THIS IS SHEWN BY THEIR MOVEMENTS AND THE SOUNDS UTTERED,' WE MUST ADMIT THAT THEY POSSESS SOME POWER OF IMAGINATION. THERE MUST BE SOMETHING SPECIAL, WHICH CAUSES DOGS TO HOWL IN THE NIGHT, AND ESPECIALLY DURING MOONLIGHT, IN THAT REMARKABLE AND MELANCHOLY MANNER CALLED BAYING.

Darwin was, of course, indulging in a piece of speculation. More recent studies have shown that at least the higher primates display REM while sleeping, and therefore probably do dream. As to what their, and our, dreams mean – that is a matter of sheer guesswork. As I have written in another context (*Dreamers*, 1984), 'Humanity is going to look pretty silly if it turns out that dreams don't mean anything at all.'

Do Human Beings Have an Innate Sense of 'Place'?

We like some places and dislike others for reasons which are hard to understand. The obvious explanation of this is that we have learned subliminally that certain factors, in combination, indicate that one place is safe for us while another is not. However, obvious explanations are not always correct, and some studies carried out during the past couple of decades seem to suggest that, instead, we are born with an inherited knowledge of the signals that can indicate danger. Even the youngest of babies will automatically avoid precipitous drops in floor-level – and, perhaps more interestingly, places where it *looks* as if there is a precipitous drop in floor-level. There is little evidence that the babies' caution has been learned by example from their parents; it seems more likely that the babies have inherited the knowledge of the relevant danger signals.

Some scientists have taken these ideas a lot further. For example, in a 1975 study Jay Appleton wrote that people inherit a preference for places that are

- a prospect – they allow a good view of the surrounding territory
- a refuge – the individual is invisible to people in the surrounding territory

Clearly one of the ideal places to be, so far as Appleton's theory is concerned, is in the topmost branches of a leafy tree. Since our ancestors presumably sought safety in exactly such spots, his theory would seem to be plausible – assuming that acquired information can be transmitted genetically from adult to offspring, which is unlikely. Another scientist, Stephen Kaplan, basing his work on that of earlier researchers,

ABOVE *Thick rime on a yacht moored in the harbour of Geneva, Switzerland. Most people feel chilly instinctively when looking at a picture like this, even though the photograph itself is, obviously, not cold. We are responding to the visual cues.*

has presented a different version of the same idea. He proposed that humans innately prefer places that offer both

- mystery and complexity – they interest us and give us the promise that we can delve to find out more about them
- coherence and legibility – we understand them, and therefore feel safe in them

Kaplan's theory would seem to be the more plausible of the two, in that it does not depend on parents transmitting the results of their learning to their children; instead, it can be explained purely in terms of human curiosity. However, it shares one great difficulty with Appleton's idea: it is impossible to test. We can see this if we take an example. Many people find the night-time skyline of a big city very exciting: they enjoy surveying it, and are content to be surrounded by it. Might this be solely because its mystery and complexity are recognized at an instinctive level? One way of trying to test this might be to show the skyline to very young babies and see how they responded. The difficulty is that, in order to give the results of the tests any meaning, we would then have to show other – less or more complex – skylines to the babies, and see how they reacted to those. But how could we ensure that complexity and mystery were the *only* qualitative differences between the various skylines?

It is certainly possible that we do indeed have an innate sense of 'place'. For the moment, however, we have to regard the matter as unproven.

What Determines a Child's First Words?

LEFT *Sunset in the Seychelles, a scene of tranquillity that inspires tranquillity in the viewer.* ABOVE *Another scene in the Seychelles, this time offering us tranquillity for very different reasons; the rocks and the overshadowing branches indicate to us that here we can enjoy good observation of our surrounds without ourselves being observed.*

One of the oddities of child development is that the word meaning 'mother' is so similar in so many languages. That the Sanskrit *matr* should give rise to the Greek *meter* and hence the Latin *mater* may seem hardly a surprise, since each of these languages influenced its successor. What is less easy to explain is that all of these words seem to have been created as phonetic imitations of the first meaningful sound uttered by many babies. This sound begins with the enunciation of the letter 'm', and can be variously 'mama', 'mummy', 'maman', 'mutta', 'maaa' and so forth; the feature of the baby's environment designated by the word is almost invariably the mother. The reason for this trans-cultural similarity of language is not known; it may simply be that 'm' is the easiest of all the consonants to produce.

Some babies ignore the 'rules'. They may decline for some quite considerable time to use any particular sound to indicate their mother, instead preferring to fix on some other aspect of their environment. My own child's first 'word' was 'doidledoidledaddy', a fact that rightly infuriated her mother. Other children can pick on such ephemera as 'toaster', 'video' and 'potty'.

This points up a second mystery. It is easy enough to accept that a child's first word should label its mother, until we start to think about it: why should a child connect sounds with objects at all? There is no obvious reason. After all, the relationship is a fairly remote one: the young child equates a sequence of consonants and vowels with a physical object, which is by no means an obvious connection. The baby could simply point, which is what quite a number of them do.

The riddles of babies' first words have yet to be solved. At present little research is being done on the subject.

How Common Is Cannibalism?

The eating of human flesh and the drinking of human blood in order to absorb the qualities of the deceased are practices which seem to date back as far as humankind itself. The Judaeo-Christian tradition regards cannibalism as a sin, but other major religions beg to disagree, on the grounds that, if a person is dead already, the consumption of his or her body is not going to make much difference. During the Middle Ages Christian countries took this idea one step further: drinking the blood of a just-beheaded criminal was widely regarded as an effective cure for epilepsy, and mixtures containing powdered human bones and/or flesh were popularly regarded as aphrodisiacs.

That drinking human blood can increase fertility is an old idea. It is related that Annia Galeria Faustina, the wife of the Roman Emperor Marcus Aurelius (121–180), was so desperate to become pregnant that she drank the warm blood of a dead gladiator. Her son became the Emperor Commodus, one of the nastiest Roman rulers of all. Clearly it would be unscientific to make any connection between Annia's and Commodus's different versions of bloodthirstiness!

The practice of eating dead enemies probably arose from the idea that you could absorb the better qualities of the person who had died through eating their flesh. Conversely, eating the person's body might indicate the ultimate contempt, in that you were degrading them completely: as you ate parts of the body you were mocking the person's erstwhile vigour and simultaneously stealing it for your own use. A couple of examples are worth noting. As recently as 1971 a member of the Black September organization boasted proudly of drinking the blood of the assassinated Wasfi Tal. Later in the 1970s Idi Amin, then dictator of Uganda, was

LEFT AND RIGHT *Two famous figures reputed to have indulged in cannibalism, Annia Galeria Faustina and (pointing) Idi Amin, one-time dictator of Uganda.*

accused of eating parts of the human beings who, indubitably, died in his torture camps.

Did any of this happen?

The evidence is remarkably elusive. A 1979 article ('The Man-Eating Myth') by Professor W Arens, surveys a wide range of cannibalism stories concerning the Caribbean, South America, New Guinea and West Africa, and finds that every account is based on hearsay rather than eye-witness reports. His conclusion is that tales of cannibalism have yet to be proved – they are travellers' tales rather than anything else. Even if we agree with his general hypothesis, we have to accept that in some instances people will eat the dead bodies of other people – if only in order to stay alive.

The conclusion to which we can come is that cannnibalism is rare but that, *in extremis*, it is something to which the human species will resort.

Why Are Human Beings Monogamous?

Humans are not the only animals to display the habit known as pair-bonding – the practice whereby parents stay together throughout the time that their offspring require to attain adulthood. Yet there is something of a mystery in this connection. Human males are capable of fertilizing almost as many human females as they would wish; some of our close relations among the higher primates do exactly that. Among the primates it seems that the identity of the mother is important; that of the father is irrelevant. (The role of males in baboon tribes provides an interesting exception to the general rule.) Yet human parents generally stay together until their offspring are fully grown – and may even cohabit until the end of their lives despite the fact that they bear no more offspring together.

The reasons for this are very far from clear. In part they seem to be societal – you score no points if you ditch your spouse – but largely they seem to concern the very basic matter of pair-bonding. This ignores the fact that some men and some women wish to encounter as many sexual partners as possible, behaviour which may have good biological reasons behind it. It seems unlikely that monogamy is a natural state for the human animal.

PART THREE

Mysteries Of The Universe

All mysteries of science are, by definition, mysteries of the Universe, which, for the sake of argument, we can say is the sum total of all matter, all energy, and all events that have taken place throughout past time and will take place throughout future time. In this part of the book we shall consider mysteries on the grand scale – and also on the inconceivably tiny. At one moment we shall be talking in terms of billions of years and billions of light years; at the next we shall focus on events that are of importance for only billionths of a second involving particles so small that it is misleading to think of them as material objects.

There is no paradox in this. Those tiny events and particles mould the Universe as a whole. Stars could not shine were it not for particles – called neutrinos – that are so insubstantial that they have no mass at all and can travel right through a solid object like the Earth as if it simply did not exist. And the birth of the Universe, if some current theories are to be believed, depended on particles that did not have any physical existence at all.

LEFT *Fred Hoyle in 1955, some years before he, Bondi and Gold propounded the Steady State Theory of the Universe. Hoyle, still active today, has never been scared of iconoclasm.*

How Did the Universe Come into Existence?

A few decades ago this would have been a much more controversial question than it is today, because one possible answer would have been that the Universe had always existed, and always would. This was because of a cosmological theory that was then very popular – the Steady State Theory propounded by the US scientist Thomas Gold and the UK scientists Hermann Bondi and Fred Hoyle.

One of the conundra of the Universe is that galaxies – vast islands composed of millions or billions of stars – are almost without exception receding from each other. We can tell this because the light from their stars is reddened, in the same way and for the same reasons as the noise of an ambulance siren or a car engine seems to change pitch downwards as the vehicle passes you. In whatever direction we look, the

LEFT Radiotelescopy has allowed us to examine the most remote areas of the Universe and determine the details of celestial objects that are so distant that they are invisible to optical telescopes. This is one of the instruments at the Nuffield Radio Astronomy Observatories, Jodrell Bank, England.

RIGHT *The pioneer of radioastronomy, Karl Jansky, with his directional radio aerial system, the precursor of modern radiotelescopes.* **OPPOSITE PAGE** *Antimatter particles were predicted before there was any hard evidence that they existed. In this photograph from CERN we see the annihilation of an antiproton in the 80cm Saday liquid-hydrogen bubble chamber.*

light from the galaxies shows this redshift; moreover, the more distant a galaxy, the greater its redshift. Some of the nearest galaxies are moving very slowly towards us (that is, the light from them shows a blueshift) but that is a purely local effect. The clear inference is that the Universe as a whole is expanding.

The obvious conclusion is that, at some moment in the far past, the Universe was compacted into a very small volume – the 'cosmic egg' – and then exploded in a 'Big Bang'. Gold, Bondi and Hoyle disagreed. They were reluctant to believe that there could ever have been a time when the Universe had not looked much like it is today. Also, the Big Bang Theory implied that the Universe had a finite extent – it had an 'outside' as well as an inside – and this was another idea that they could not stomach. They therefore proposed the idea of *continuous creation*, whereby matter was constantly popping into existence. The emergence of this matter would be sufficient to, as it were, 'push apart' the galaxies. Not very much new matter was required: Hoyle pointed out that, throughout the Universe, the appearance of a single hydrogen atom each century in a volume equal to that of the Empire State Building would be quite enough.

Unfortunately for the Steady State Theory, in 1965 two US electronics experts, Arno Penzias and Robert Wilson, detected a residual background microwave radiation present throughout the Universe. The only possible explanation of this radiation, which corresponds to a temperature of about 3K (3C° above the absolute zero of cold), seemed to be that it represented the residual energy left over from the Big Bang. The Steady State Theory floundered on for a few more years, being continually revised, but eventually its three proponents conceded defeat.

However, the idea of the 'cosmic egg' was in itself not very satisfactory. If the 'egg' had existed for all eternity, why should it suddenly decide to explode? Conversely, if the egg had just popped into existence and then exploded, what could explain this bizarre event? Clearly there were as many flaws in the current version of the Big Bang Theory as there had been in the Steady State Theory. Another question troubled cosmologists' minds. Was the Big Bang a unique event? Was the Universe destined to go on expanding forever, or might it be that gravity would eventually pull all the galaxies back together again until there was a 'Big Crunch' – later to be followed by a new Big Bang? This mystery has yet to be solved, although two quite separate sets of theories seem to indicate that indeed there are successive Big Bangs.

Our modern ideas of what caused the Big Bang seem almost mystical. In order to grope towards an understanding of them we have to grasp the abstruse notion of the *particle sea*. This idea was born out of the theories of the brilliant British physicist Paul Dirac, who as long ago as 1930 proposed that there must be an analogue of matter called *antimatter*. To take a simple example, the important subatomic particle called the electron has a negative electrical charge. Dirac worked out mathematical equations that indicated that there ought to be a counterpart to the electron but having a positive electrical charge. Soon afterwards exactly such a particle – the antielectron, or positron – was detected, and we now know that for every type of particle there is an appropriate *antiparticle*.

In the same way that matter is constituted of fundamental particles, antiparticles are the building blocks of antimatter. However, matter and antimatter annihilate each other completely when they come into contact: an 'antimatter bomb' would create an explosion giving off such stupendous amounts of energy that the H-bomb would pale by comparison. It is thought possible that, when the Universe came into existence, there was by chance a little more matter than antimatter; most of the matter destroyed itself by interaction with the antimatter, but that still left some over – the matter which makes up the Universe around us. (It is possible that there may be areas of the Universe where antimatter galaxies, stars, planets and even life exist, but we have no evidence of this.)

How do these ideas relate to that of the particle sea? The suggestion is that the whole of the Universe is filled with *virtual particles*. In the simplest terms, these can be thought of as particles which at the moment do not in fact exist – but one day might. Just as a particle and an antiparticle will totally annihilate each other if placed in contact, there is no reason why pairs of particles and antiparticles cannot abruptly spring into existence – the accent being on the word 'pairs'. Because the particle and antiparticle in effect cancel each other out, their appearance together does not violate the laws of physics. Of course, almost always the two immediately annihilate each other, so the event might never have happened, but it is possible that on occasion, for one reason or another, the two are separated so swiftly that annihilation does not occur. Experiments have shown that these virtual particles are indeed present, although we cannot detect them directly.

We can imagine, then, a circumstance in which there were no such things as matter, antimatter or energy – only a sea of virtual particles. There were would be no such thing as *time*, either, in this situation, because for time to exist there must be events. All that is needed to disrupt this state of affairs is for a 'seed' to be planted. Perhaps some particles spontaneously emerged. The effect would spread like wildfire – but with a ferocity and energy release that no wildfire could ever match. There would be a colossal explosion – in other words, a Big Bang.

It must be stressed that this is only a possibility (and that the explanation has been very much simplified), but more and more researches are showing that it probably approximates to the truth. The precise circumstances of the birth of the Universe, perhaps some 15 billion years ago, are still a mystery – and are likely to remain so for many years.

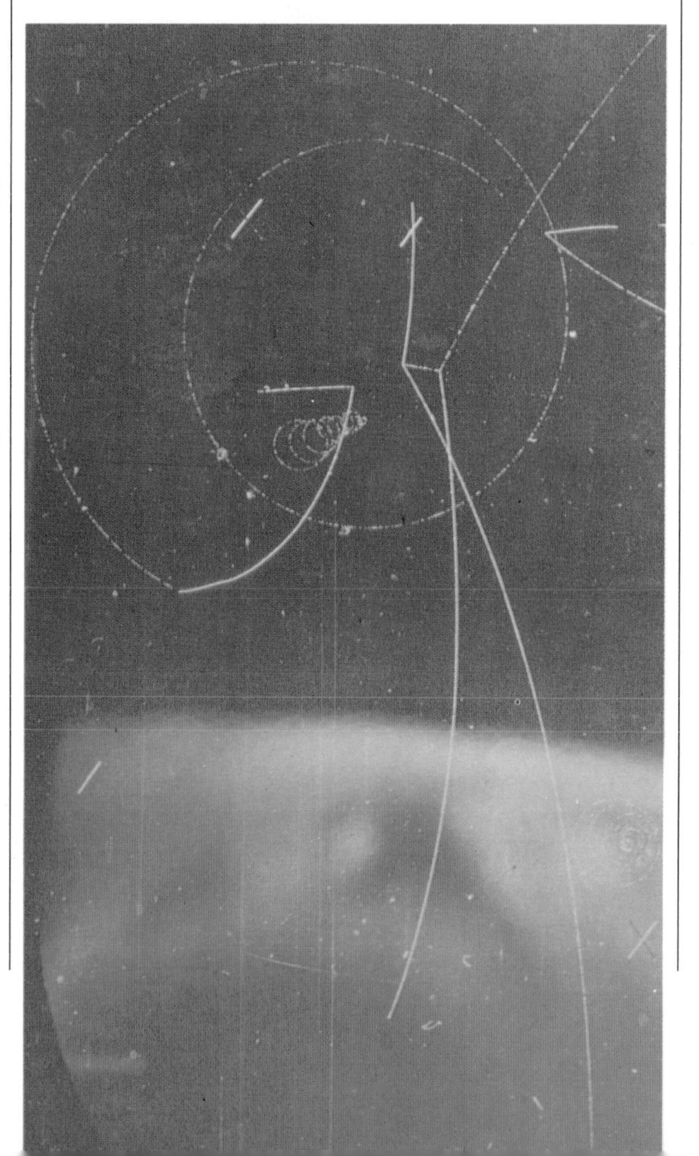

Could There be Many 'Little Bangs'?

According to the accepted theory, then, there was no such thing as time before the Big Bang. The explosion was not a simple three-dimensional one. Rather, it occurred in four dimensions: the three to which we are accustomed, plus time. In fact, the event was even more profound than that, because it was also an explosion *of* the four dimensions (as well as, it is thought, a number of others, which almost immediately 'rolled themselves up' so that we do not recognize them as such) – the matter and energy involved being only a secondary consideration! In a very real sense, then, the Big Bang Universe has existed forever.

Some scientists – and we should add hastily that they are in a small minority – feel that this is mincing words: they feel that by 'forever' we should mean an infinite number of billions of years rather than a period whose duration can be estimated. They are unhappy with the idea of a Universe that has not existed infinitely looking much the same as it does today – the same consideration which fuelled the Steady State Theory. This leaves the problem of how to explain the microwave background radiation.

One suggestion has been that perhaps the matter and energy of the Universe are brought into existence by an infinite series of 'Little Bangs'. These are events similar to the Big Bang but on a far smaller scale. Their cumulative effect would give rise to the microwave background radiation. The idea can also be used to explain the fact that the galaxies are all receding from each other because, just like the Big Bang, the 'Little Bangs' create not space (and trivia like matter and energy) but *spacetime*.

Here we need to pause to think about what we mean by spacetime. We all know that we exist in three dimensions – our bodies occupy a certain volume, and we can move it from one place to another. We know, too, that we exist in the fourth dimension, time, because we live for a certain number of years; besides, the fact that we can move in three dimensions implies our existence in the fourth, because movement requires time. That we think of this fourth-dimensional aspect of our lives as being in a different 'category' from the rest is really a matter of perception rather than reality. In order to define completely where you are right now you would need to use four mathematical coordinates, not three, and each of these would have exactly the same status as the rest. In terms of where you are relative to the Earth's surface, the four coordinates could be:

- latitude
- longitude
- altitude
- time

In short, we can see that we do not exist separately in space and time but in spacetime – and the same is true of everything else in the Universe.

Discussions involving spacetime inevitably turn to three-dimensional analogies: it should always be borne in mind that these are simplifications. Let us take a balloon as our model of the Universe, and its surface as representing spacetime. As we blow up the balloon, its surface gets larger (that is, spacetime expands). If we glued little pictures of galaxies to the surface, we would see that they moved progressively further apart as we inflated the balloon. The energy source powering this expansion is, of course, our breath.

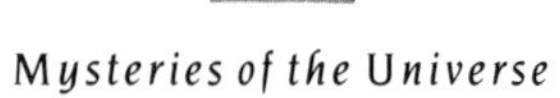

TOP *The cluster of galaxies in Fornax.* **ABOVE** *The spiral galaxy NGC 6946. Here we can see both spiral and elliptical galaxies.* **LEFT** *The core of the nearest major galaxy to our own, the Andromeda Galaxy M31.*

LEFT *The Sombrero Hat Galaxy, one of the prettiest spirals.*

But there are two ways of blowing up a balloon. The ideal way is with a single, long exhalation, but most of us do not have the lung-power for that. We use a succession of shorter puffs – and here we get back to the 'Little Bangs'. The expansion of spacetime can be explained just as reasonably by the effects of a series of 'Little Bangs' as by that of a single Big Bang. The Universe would therefore have an infinite extent in all four dimensions of spacetime; it would always have existed, rather than having a moment of birth.

This is all very well, but even a 'Little Bang' should be a pretty spectacular affair. Yet, if we try to find one, we search the heavens in vain. For a while supporters of the theory placed great faith in quasars – very distant galaxies whose cores are giving off colossal amounts of energy. However, we now know that quasars are not 'Little Bangs' in progress.

Of course, 'Little Bangs' would necessarily be widely separated, so it is perfectly possible that there simply aren't any near enough to us to be detected: in an infinite Universe, our local patch – vast though that volume of space might seem to us to be – would be very small indeed.

The 'Little Bang' Theory cannot be completely dismissed, but it is today generally discounted. The problem is that, the further one goes into it, the more involved the science and mathematics become, with new rules and caveats having to be brought in to explain away apparent discrepancies. The Big Bang Theory, which is anyway much simpler, explains phenomena rather than, conversely, requiring postulated phenomena in order to explain it, and therefore must be for the moment preferred – although elsewhere we shall consider a variant of the 'Little Bang' Theory.

Are Black Holes Gateways to the Universe?

Black holes have had tremendous press coverage in recent years, so there can be few who have not heard the term. However, there is a great deal of public confusion about the nature of black holes: only a few years ago a television serial depicted astronauts landing on the surface of one!

Stars evolve. Their evolution is governed by the nuclear reactions going on inside the star and making it shine. For much of its life a star shines because of reactions in its core which convert hydrogen, the lightest of all the elements, into helium, the second lightest. Eventually, however, the hydrogen in the core runs out and so nuclear reactions involving heavier elements take over. The core contracts under the weight of the star and the hydrogen-fusion reaction continues in regions progressively further away from the core. Because

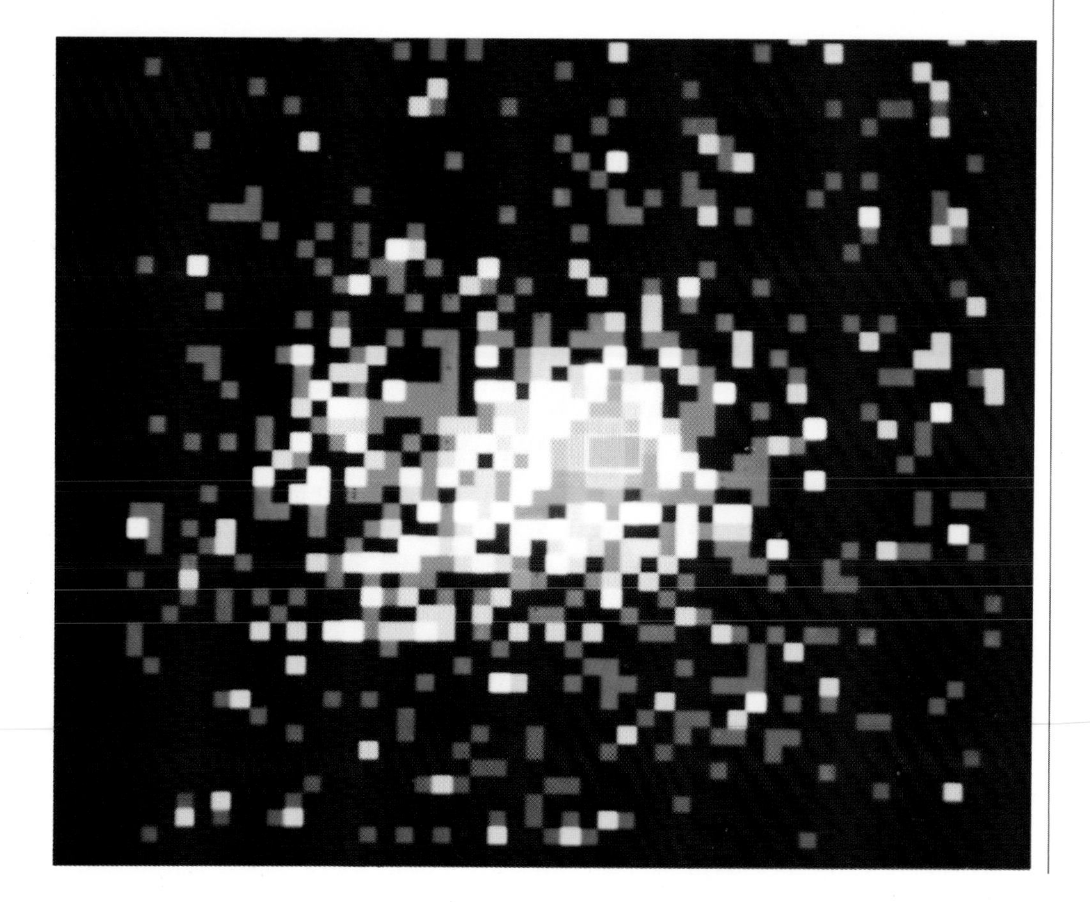

RIGHT *Black holes are the likely outcome of supernovae, such as the one pictured here, the brightest one to be discovered since 1604. The image is colour-coded and digitized by computer; red represents the brightest parts of the image, fading to green, blue, white, grey and black. The supernova is visible as the cluster of red and blue pixels (squares) in the centre of the image.*

of the internal pressure of these reactions, the star swells up to many times its former size to become a 'red giant'.

However, this cannot go on forever: there comes a stage when nuclear fusion to produce yet heavier elements *requires* energy, rather than producing it. The star runs out of fuel.

What happens next depends on the star's mass. A modest star like our own Sun will contract under its own gravity in a fairly orderly fashion to become a comparatively small object known as a white dwarf. Over billions of years it will slowly fizzle out to become a black dwarf.

A rather larger star – one of mass greater than about 1.2–1.4 times our Sun's – cannot contract in such a controlled way. Instead it collapses in on itself catastrophically to produce an object a few kilometres across, known as either a neutron star or a pulsar; often this is accompanied by a stupendous explosion, a supernova. (To understand the violence of this explosion, we can realize that a supernova may temporarily shine brighter than all the rest of the 100 billion or so stars in its galaxy put together.) In a pulsar, all the matter has been completely crushed: it no longer consists of atoms, because all the protons and electrons have been jammed together to form neutrons. A piece of this material, neutronium, the size of a pinhead might weigh a billion tonnes or more (depending upon the size of the pinhead!).

A worse fate awaits even larger stars – those initially of about 10 or more times the Sun's mass. The force of their own gravity is so great that their collapse cannot be halted at all. The matter of which they are made is crushed right out of existence . . . but it leaves behind it a gravitational field, as if it were still there but compressed into a very small volume of space. When we talk about black holes, what we are really referring to are these gravitational fields.

It is difficult to think about a gravitational field that exists independently of any associated matter. After all, the Earth pulls us because it is *there*; likewise, the Sun keeps the Earth in its orbit because it is *there*. We can get some idea of what is going on by thinking of a sheet of canvas stretched on a horizontal frame. If we push down on the canvas by putting a heavy object on it we create a dent. If we then roll a ball towards this dent, the ball will swerve as it goes by. In an ideal situation, the ball might approach the dent in such a way that it continued to go round and round it in a circle – rather like the Earth goes round the Sun. Now imagine that we used a ridiculously heavy weight, so that it broke right through the canvas, leaving the dent in the stiff cloth but with nothing at its centre. When we now roll the ball towards the dent it may circle around once or twice but eventually it will fall through the hole. The ball behaves initially as if the heavy weight were still there; but after that its fate is sealed.

Black holes are very small indeed. If we could somehow persuade our planet to become a black hole, it would be about the size of a marble – but still possessing all the mass of the original Earth. (Conversely, if we compressed the Earth until it was the size of a marble, it would automatically become a black hole.) At the 'surface' of this black hole the pull of gravity would be about 1,600 million billion times as great as the pull we normally experience.

ABOVE *'Preconstruction' by Andrew Farmer showing a hypothetical probe approaching a black hole that is part of a binary system. The hole's gravity is stripping away surface material from its companion star to create a typical accretion disc.*

The size of a black hole is determined by its escape velocity. We are used to the idea of escape velocity when talking about spacecraft taking off from Earth: unless they can achieve a velocity of about 11.2km (18 miles) per second they will never achieve orbit. The velocity of light, the greatest possible velocity, is about 300,000km (186,000 miles) per second. The 'surface' of a black hole is therefore the place where the escape velocity is that of light. In other words, nothing, not even light, can escape from beneath this 'surface'. Not for nothing are black holes called black.

In fact, because of curious effects connected with the quantum theory and involving obscure mathematics, energy *can* gradually seep away from black holes until eventually they disintegrate. So far as we are concerned, this barely matters: although it is likely that 'mini black holes', about the size of a subatomic particle, were created by the enormous forces involved in the Big Bang, it is only about now, perhaps 15 billion years later, that these should be beginning to disintegrate. Larger black holes will take indefinitely longer.

Black holes have been seen as a nemesis. Like cosmic vacuum cleaners, they must sweep into themselves matter and energy from all around them. As they grow they become more powerful vacuum cleaners, affecting a larger and larger volume of space, and themselves becoming ever larger. Since the lifetime of a large black hole must be measured in untold billions of years, it seems inevitable that the end of the Universe can be nothing else than the coalescence of everything into a single black hole.

Yet a question remains. If matter and energy are crushed completely out of existence inside a black hole, where do they actually go to? The idea that they vanish completely from existence seems to offend against logic, yet they cannot escape from the black hole. One answer that has been proposed is that they somehow 'tunnel' their way instantaneously to another part of the Universe – or even to an alternate Universe. Oddly enough, the mathematics of the situation reinforce this possibility. In theory, then, we could use black holes as a way of travelling not just between the stars but between the galaxies.

There are some problems with this optimistic scenario, however. First, in the case of the ideal black hole we have been talking about, our spaceship and its crew might indeed travel to the far end of the Universe – but in the form of a spray of raw energy. Still, in reality no black hole would be like this ideal: stars spin on their axes, and so the black holes they produce can be expected to do likewise. The mathematics for a spinning black hole are rather different from those of a stationary one, and initially it was thought that they would allow the possibility of a suitably durable craft surviving the voyage without necessarily being destroyed. Even so, navigation would be a problem – especially since you could not predict where you would arrive. Even worse, you could not predict *when*, since you could find yourself

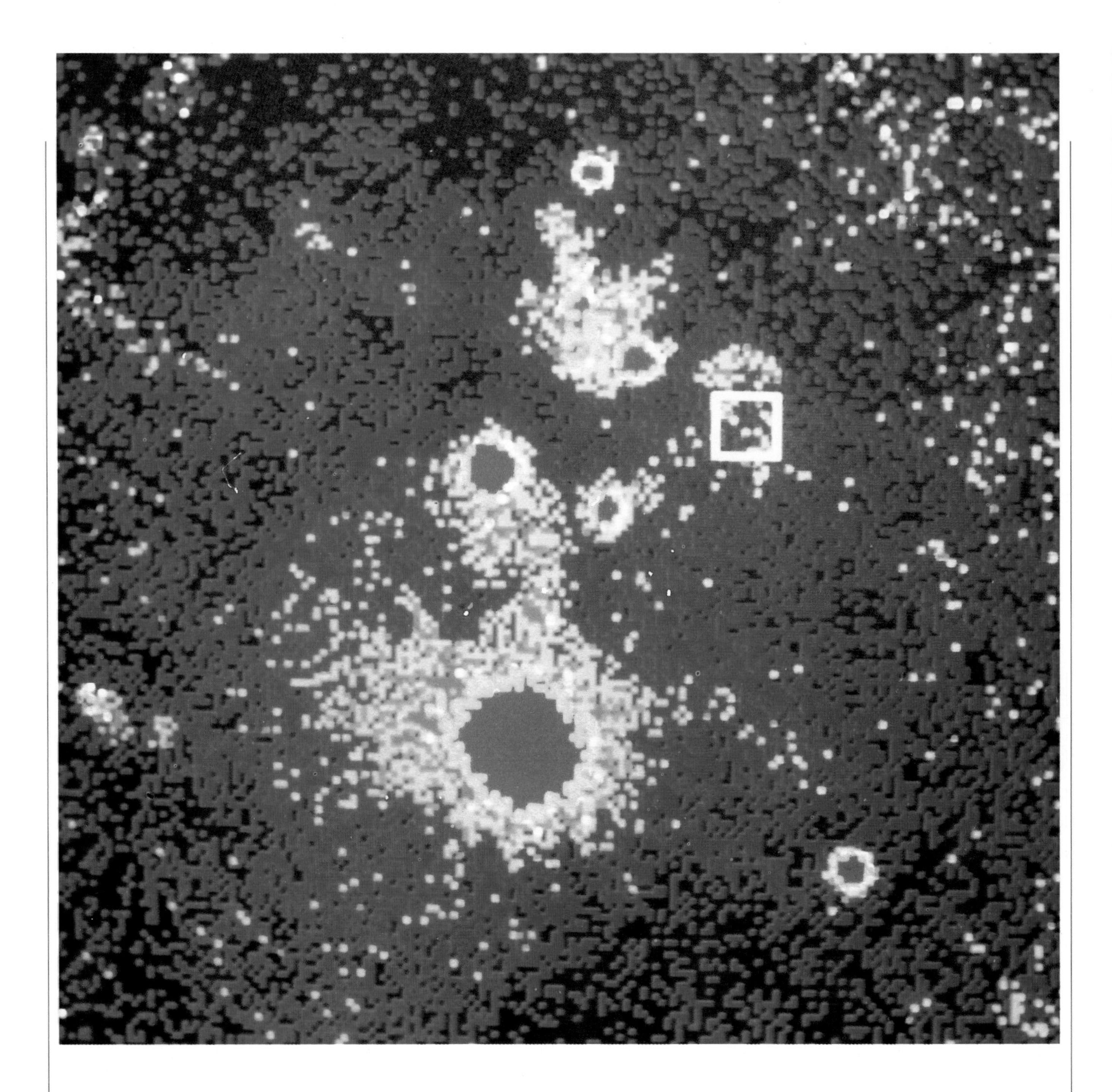

ABOVE *The supernova 1987a photographed in ultraviolet light by* Explorer. **OPPOSITE PAGE** Quasar 3C273. *The jet of matter being expelled is partially visible in the light spectrum.*

re-emerging at any time, past or present. Navigation would be even more difficult if you tried to come home again – but you could well return before you set off. It is perhaps fortunate for our mental stability that more recent researches seem to indicate that black-hole travel is impossible.

Black holes could form a useful means of travel into the future. According to the Theory of Relativity, time passes more slowly for objects moving very swiftly; thus a spacecraft moving at close to the speed of light would take more than 100,000 years to cross our Galaxy, as observed by the rest of us, but only a few years so far as its crew were concerned. Gravity has a similar effect; a spaceship could skip close to a black hole for what would seem to be a short period of time and then emerge to find that, for the rest of the Universe, millions of years had passed.

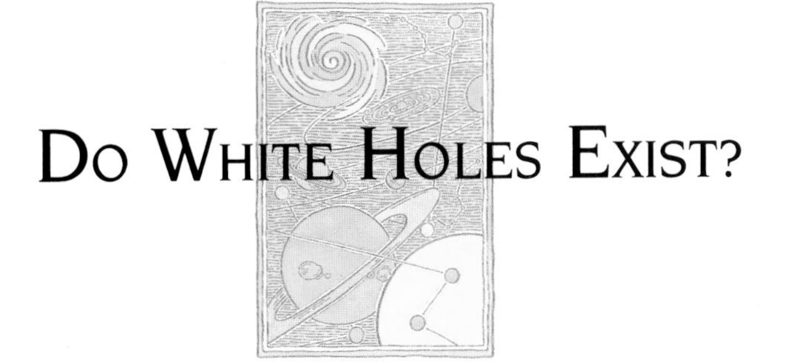

Do White Holes Exist?

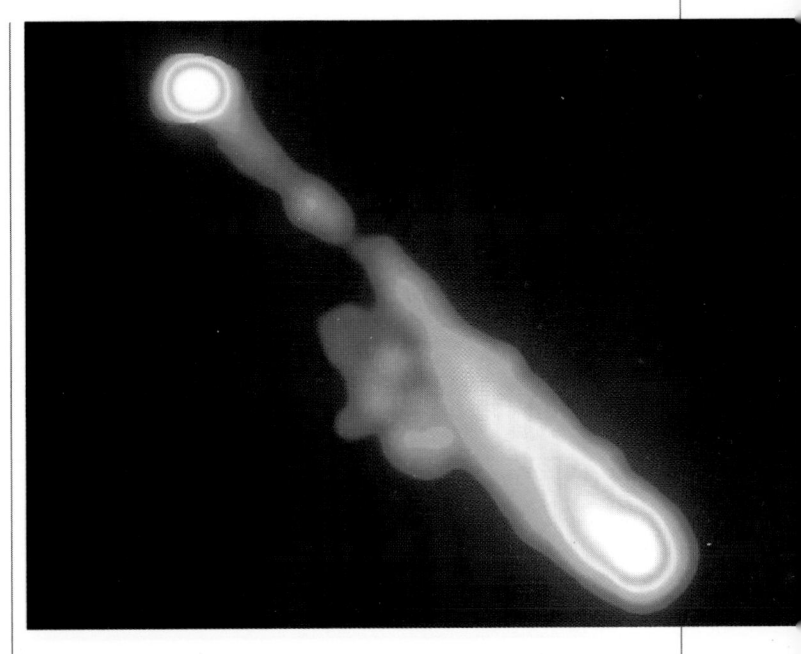

If – and it is a very big 'if' – the matter sucked into a black hole must reappear somewhere and somewhen else, we must ask ourselves what the other end of the 'tunnel' would be like. Assuming that the 'outlet' is localized, it would be a site where raw energy gushed out into the Universe as if from nowhere. This energy would be in a complete mixture of forms, from long radio waves through light to highly energetic X-rays. Some of it might even be, at least briefly, in the form of matter, since matter and energy are interchangeable. Such a hypothetical 'outlet' is called a white hole, and in simplest terms is the opposite of a black hole.

If white holes exist they would seem to provide, in conjunction with their corresponding black holes, a rather neat way for the Universe constantly to replenish itself. Anything lost into a black hole would, at some time or another, be returned via a white hole. Moreover, white holes, by acting as conduits to bring new material and spacetime into new regions of the Universe, would act in rather the same way as 'Little Bangs', at least on the local level (the 'locality', of course, being extremely large). The influx of energy would help explain the microwave background radiation.

However, we have the same problem as with 'Little Bangs': white holes should be extremely obvious cosmic phenomena, broadcasting powerfully in every region of the energy spectrum, and yet we can find no trace of them. People have looked with interest but in vain at the cores of quasars, but they are certainly not white holes. In short, we still have no direct evidence that white holes exist.

There is a conundrum here. It is as certain as most things in astronomy can be that black holes exist – and are, indeed, common. They are predicted by theoretical physics, and there is good experimental evidence in favour of their existence. But the same theoretical physics predicts white holes, and yet we have no experimental evidence in favour of what should be far more obvious objects. Could it be that the Universe's first white holes have yet to form? Could it be that our assumption that matter and energy are returned to the Universe in discrete regions is wrong, that in fact they are returned piecemeal all over the Universe? Or could it simply be that the theory is grossly at fault?

This last might seem to be the most likely possibility but, if so, we are presented with yet another mystery: how can a theory which works so well for black holes not work equally well for their direct counterparts, white holes?

Is There Such a Thing as Antigravity?

Earlier we talked about antimatter – particles which have all their physical characteristics opposite to those displayed by particles of matter. One of the properties of particles of matter is that they influence each other. There are four known forces exerted between particles of matter: the electromagnetic force (which is what makes magnets attract pins, but is more generally important on the subatomic scale), gravity (negligible on the small scale but supremely important on the large scale) and two others, the weak and the strong nuclear forces (neither of which need concern us here). A question arises: if antimatter particles have other properties directly opposed to those of matter particles, surely there must be such a thing as antigravity?

For decades antigravity has been the science-fiction writer's dream. In story after story people (or aliens) have simply switched on the antigravity drive in order to be able to travel cheaply and effortlessly between the stars. Similarly, in his novel *First Men in the Moon* (1901), HG Wells fantasized about a substance called cavorite, which could shield a spacecraft from the pull of gravity, thus allowing its propulsion units to work *very* much more efficiently!

Some of the 'unorthodox' ideas about antigravity are fun. A story frequently recounted in 'unorthodox' circles is that of a South African called Basil Van den Berg. At some point during the late 1950s or early 1960s he deciphered – thanks to the help of a passing visitor from the planet Venus – some hieroglyphs from the Amazon Basin which were dated to about 10,000 years ago. As you might expect, the message of these hieroglyphs was nothing less than the instructions as to how to build an antigravity propulsion unit (based on magnetism), like the ones which the Venusians had used in their spacefaring days. Van den Berg built a prototype ... which presumably worked, because neither hide nor hair of him has ever been seen since.

From the ridiculous we can turn to the sublime. Gravity attracts objects towards each other; antigravity would make them repel each other. How this could work depends on the nature of the objects concerned. Two antiparticles could therefore be drawn together by antigravity, but a particle of matter would be repelled by an antiparticle. Gather together enough antimatter, and you have a cheap way of travelling to the stars – because obviously the only energy you would be using would be gravitational, of which there is plenty available in the Universe. Unfortunately, it seems extremely unlikely that predictable technology will be equipped to amass large quantities of antimatter.

Yet the discussion is not without interest. As we have seen, it is probable that pairs of virtual particles – one of matter, one of antimatter – spontaneously pop into existence all over the Universe. Normally they do so in such intimate contact that they immediately annihilate each other, because the force of gravity – and hence of antigravity – is weak over tiny distances when compared with the strong nuclear force and the electromagnetic force, which pull the particle and antiparticle inexorably together. On the fringes of a black hole, however, all this could change. The particle of matter is instantly whipped towards the black hole by the huge gravi-

RIGHT *Quasar 3C179, showing a classic radio-quasar structure with a double lobe, a strong core, and a one-sided radio jet that undergoes significant bending as it enters the western lobe.*

tational tug; the antiparticle is simultaneously repelled, and zips off to become a part of our Universe – although probably not for very long, because it will annihilate itself as soon as it encounters another matter particle.

This annihilation is in itself interesting, because the energy released by the explosion is equivalent to the mass of *both* particles. Since the antimatter particle did not have any existence before, the Universe has received an injection of extra energy; on the other hand, it has lost the energy of the particle that fell into the black hole, so things seem to have evened themselves out. Except, of course, for the fact that it may take some while before the matter particle falls into the black hole – in which case the Universe has a temporary and minor energy surplus. Multiply billions of tiny energy surpluses over billions of years of time and you find that you have a vast energy surplus. Perhaps this process could be the one responsible (or partially responsible) for the expansion of the Universe?

Another point concerning antigravity can be stated quite simply. Black holes are regions possessing hugely strong gravitational attraction. White holes, by contrast, are presumably areas of intensely strong antigravitational repulsion. But then, if they are in regions where antimatter dominates over matter, the repulsion would in fact be an attraction, and the white hole would be indistinguishable from what is, in our Universe of matter, a black hole.

The case for antigravity must be regarded as 'not proven' – as Scottish jurors decree when they cannot decide upon the guilt or innocence of the defendant. However, it seems unlikely that the antigravity drive of science fiction's dreams will ever be realized. On the other hand, there is just the remote possibility that someone will invent it tomorrow.

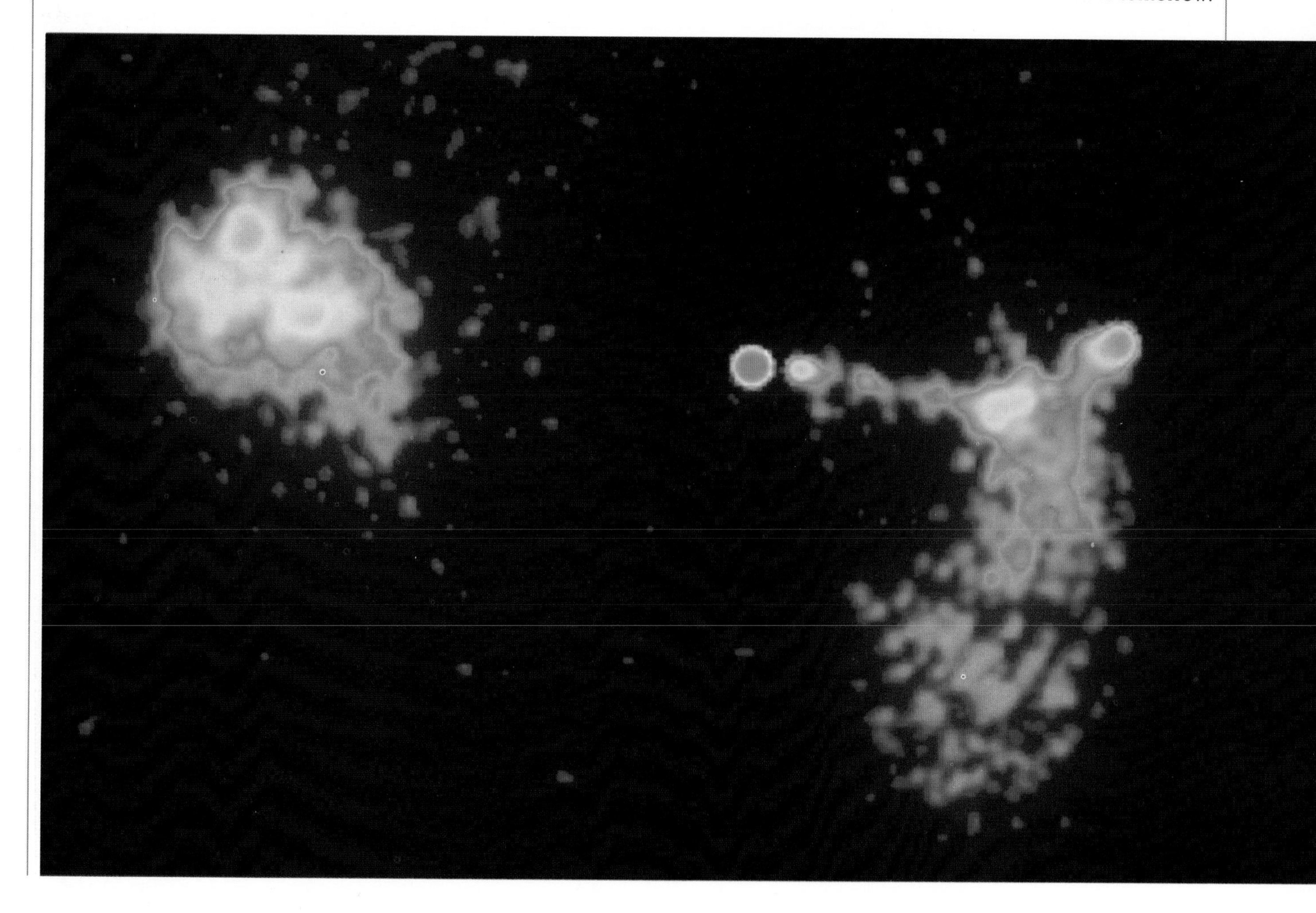

What are Quasars?

The word 'quasar' is shorthand for 'quasi-stellar object'; the longer term gives a very clear indication of what quasars look like – from Earth. In the telescope they look like very faint stars, but they have enormous redshifts, indicating that they are retreating from us extremely swiftly and are thus distant – the most distant objects of which we know. This means in turn that they must be shining almost unbelievably brightly – perhaps 100 or more times than an average galaxy. However, many of them show random variations in brightness lasting only a few months or, on occasion, days.

This is very difficult to explain. Because nothing can travel faster than light, an object that can vary its output of energy within a period of a few days can be at best a few light-days across. By contrast, the diameter of our Galaxy is about 100,000 light*years* – several million times greater. How could such a small volume produce such a colossal amount of energy? Initially it was suggested that it was impossible, and that quasars were in fact relatively nearby objects, their large speeds of retreat arising because, perhaps, they were being spat out by our Galaxy in some unknown way. Closer investigation showed this idea to be untenable – and radio telescopes soon showed that quasars had structures similar to those of more orthodox galaxies. A few scientists began to think in terms of 'Little Bangs'; later, more of them began to wonder about white holes.

What seems more plausible, in view of recent researches, is that quasars are galaxies at whose heart is a large and voracious black hole. As matter spirals into a black hole the enormous gravitational force pulls it asunder, and one consequence is the release of vast amounts of high-frequency energy (eg, X-rays). One minor piece of evidence in favour of this view of quasars is that the light from them, despite the enormous redshift, is typically bluish: the effect of the redshift is to bring high-frequency forms of radiation into the visible spectrum.

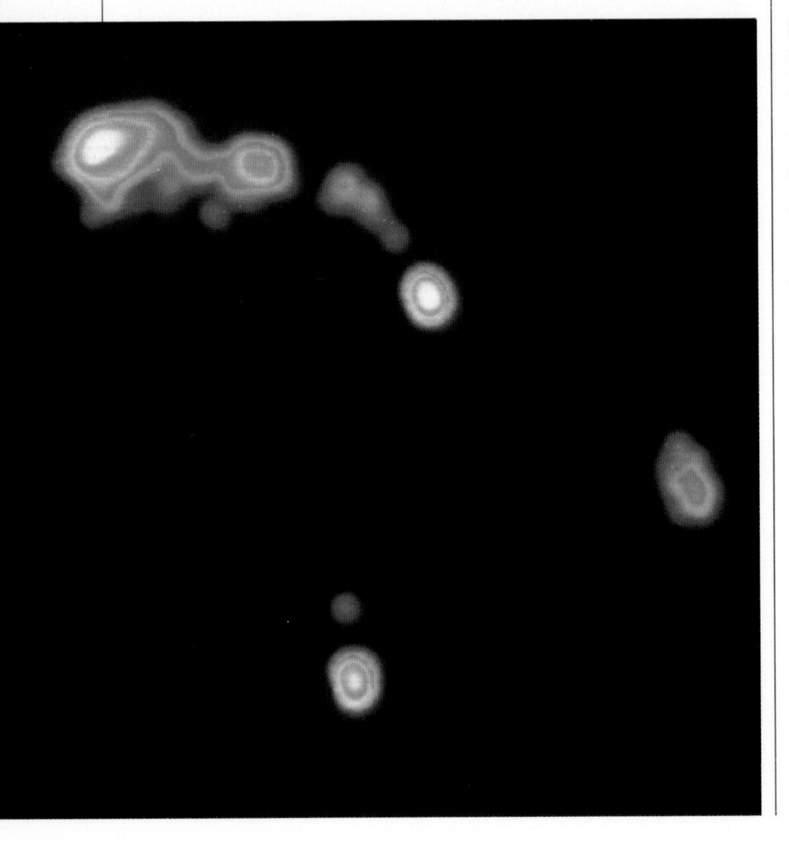

LEFT *A double image of a quasar (the two round yellowish objects). The image has been doubled through a gravitational lensing effect; that is, the gravitational field of an intervening galaxy has bent the light.* **ABOVE RIGHT** *A Seyfert galaxy,* NGC 1068.

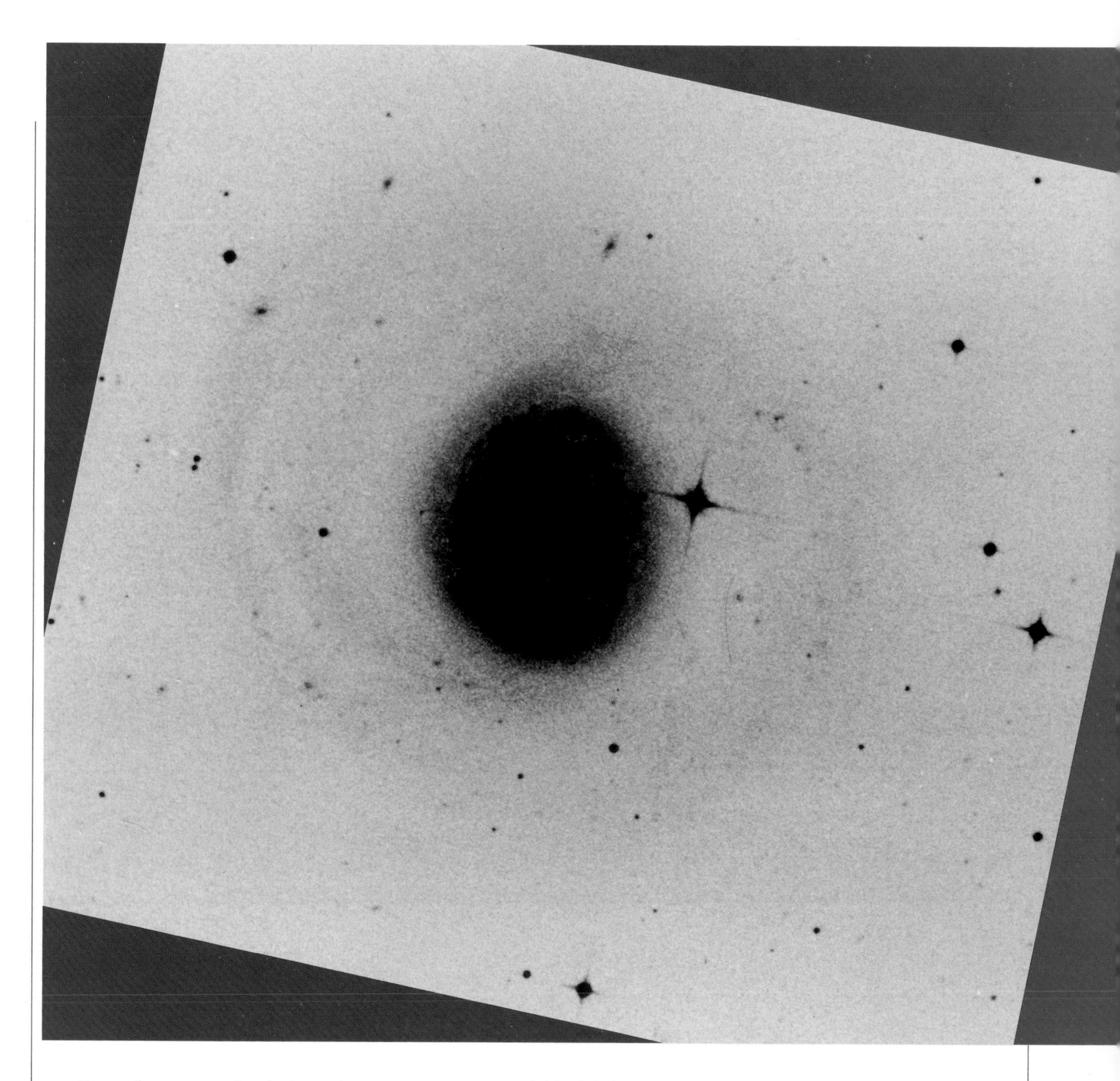

Two other types of galaxies – known as BL Lacertae objects and Seyfert galaxies – show characteristics similar to those of quasars, but 'not as much'. This has led scientists to speculate that all galaxies might go through a quasar phase, then a less active phase, and finally settle into the stability typical of our own and many local galaxies.

But black holes do not simply disappear. In other words, if these ideas are correct, then even a stable galaxy has a black hole at its core. The first frenzy of sweeping up matter is over, so that the betraying torrent of hard radiation is no longer so obvious; but nevertheless matter must still be falling into the hole. Moreover, because of the inexorable force of gravity, in due course – over many billions of years – the entire galaxy must be consumed . . . *if* these ideas are correct.

Astronomers have recently become very interested in an 'anomalous object' at the heart of our own Galaxy.

How Many Universes are There?

LEFT *The Austrian physicist Erwin Schrödinger.*

At first sight this question might seem to be a nonsense: since, as we have seen, a reasonable definition of the Universe is 'everything that has, does and will exist', surely there cannot be more than one of them? However, the definition is rooted in our own everyday ideas of reality. Unfortunately for this complacent view, science seems to indicate that there may be alternative realities – or, as they are often called, alternate universes.

The idea of an alternate universe is hard to grasp. When people first encounter it their normal assumption is that an alternate universe is nothing more than a 'what might have been'; science-fiction writers have toyed with that idea for decades. What modern science seems to suggest is something rather different: that there are other universes just as real as our own, but that there is no connection between them and our own reality. These unperceived universes could be totally separate from ours, or they could be intimately intermingled with it.

There are various reasons for thinking that this might be the case. They are as follows.

The first derives from the quantum theory. The underlying ideas are very complicated; they can perhaps be most easily visualized in terms of Schrödinger's cat. This is not a reference to any pet the Austrian physicist Erwin Schrödinger may have had but rather to a 'thought experiment' he proposed (a thought experiment is one that cannot actually be performed, but the examination of which can reveal scientific truths – or, at least, plausibilities). Schrödinger imagined an experiment in which a cat was placed in a sealed box with a sufficient supply of air to last it for the duration of the experiment. Also in the box was a bottle of poison. Over the period of the experiment there was a 50 percent chance that the bottle would be ruptured, so killing the cat. Obviously, when the experimenter opened the box, there was a 50 percent chance of the cat being alive and a 50 percent chance of it being dead. Schrödinger's question was this: during the time that the box is sealed, is the cat alive or dead? Clearly the cat will know (or not know, as the case may be), but how can anyone else tell? The answer is that no one can. Schrödinger proposed that the *reality* of the situation was that the cat was neither alive nor dead *until the experimenter opened the box to take a look*. In a sense, then, the cat's reality was not 'fixed' into either of the two options until this moment – and the experimenter was personally responible for 'fixing' it.

Schrödinger's original reason for putting forward this thought experiment was to demonstrate the folly, as he then believed, of the quantum theory: clearly it was ridiculous to think of a cat being neither dead nor alive – or, as an obvious corollary, both dead and alive at the same time. Later it was realized by Schrödinger and others that, on the tiniest subatomic scale, the duality of the cat's existence exactly represented one of the ways in which the Universe works. At any particular moment, everything in the Universe has a 'choice' – an atom of a radioactive element can decay or not decay, a photon of light can be deflected or not deflected, and so on. We can predict statistically how the atoms and photons will behave, but we cannot predict accurately for each one of them – and, afterwards, we cannot tell how each of them has behaved unless we look. Until we take that look, the behaviour of the atom or photon is not a reality: it is only one probability among many. (Moreover, the very fact of our taking that look can, as Werner Heisenberg pointed out, anyway affect the behaviour of the particle concerned. This means that there is always uncertainty in our observation of the very small: we can ascertain either the *position* or the *state of motion* of a subatomic particle, but not both simultaneously, because our observation of the particle will affect it.)

If subatomic particles are no more than probabilities until we look at them, why should any one of these probabilities have supremacy over any of the others? As the parable of Schrödinger's cat shows, there is no reason at all. Each of the probabilities – until someone comes along to 'fix' it – has equal validity, equal reality. But, of course, all the probabilities cannot be simultaneously reified in a single Universe: while it may not seem to matter much, on the cosmic scale, if an electron is in two places at once, in fact the later consequences of that tiny discrepancy might be enormous. The obvious, although unpalatable, conclusion is that there must be two universes, one for each of the electrons. But that applies to *every* electron *all* of the time – and to everything else as well, from quarks to quasars!

In other words, in each moment of time countless trillions of alternate universes are sparked off. The number of different 'yous' – all every bit as real as you are, even though they inhabit a different universe – that have been generated since you read that last sentence is literally uncountable.

This violates common sense, yet the conclusion seems inescapable. Or does it? There is another (and not generally accepted) way of looking at this situation which may seem more acceptable. When scientists draw graphs they typically plot a scattering of points which show some pattern. In order to make that pattern more obvious, the statistician then draws through the points a 'best line' – that is, a line which represents the general trend, even though some of the points lie well clear of it. Might it not be possible that ours is indeed the only Universe, and that it represents the 'best line' through all the trillions of probabilities?

This would be a comforting thought, because it would mean that all the alternate universes would be simply discarded possibilities – 'might have beens'. Less comforting, though, is the realization that this concept implies that there is no such thing as a fixed reality – which is exactly what we observe on the subatomic scale. Still, blurred or not, at least there is only *one* reality.

Unfortunately, there are other reasons to make us believe that there are alternate universes. One of these concerns the idea of the oscillating Universe (see below). Another we have already touched upon. What happens to the matter and energy sucked into a black hole? We look around for white holes and can find none. One explanation is that all the white holes are in an alternate universe. We cannot even start to conceive what conditions might be like in that universe, but there is no reason to conclude, point-blank, that it does not exist.

The existence or otherwise of alternate universes is one of the great unsolved mysteries of science – and one of those that, almost by definition, may never be solved.

How Will the Universe Die?

If the ideas of those scientists who believe in an eternal Universe are correct, then of course the Universe will never die. However, it seems more likely that our Universe

- was born at a particular moment (the Big Bang)
- is currently evolving
- will continue to evolve in the future, and therefore
- cannot exist in its current form forever

The future fate of the Universe is not one of humanity's more pressing problems: on this scale nothing drastic is likely to happen for a few hundred billion years yet. (Much more worrying is that one of the nearby stars – notably Sirius or Procyon – could explode as a supernova, bathing our planet in sufficient hard radiation to extinguish all life. This should not happen for at least a million years ... but there is an outside chance that it could happen tomorrow.) Nevertheless, if we are to try to understand how the Universe works, we should pay as much attention to its death as to its birth.

Ideas of how the Universe will die vary from one scientific theory to the next. If we assume an ever-expanding Universe in which black holes already exist, then there is no question but that, in due course, all of the matter, energy and spacetime will be swallowed by the holes. Any small amounts of matter leaking from the black holes will gather together to form new stars, but those stars will themselves evolve until they either become black holes or are subsumed into black holes. The net result is a comparatively small number of supermassive black holes – perhaps only one.

Black holes are almost immortal, but not quite. These supermassive black holes will eventually 'evaporate' through their release of radiation. This radiation will be of the lowest possible grade – in other words, heat (hence the famous term 'heat death of the Universe'). At the same time, spacetime itself will break up; as matter and energy cannot exist without spacetime as a 'basis', the end product of this process can be nothing but – nothing.

Yes, but what do we mean by 'nothing'? In fact, what we have as the end-point of the heat death of the Universe is a timeless sea of virtual particles. This may seem familiar – as indeed it should, because we have discussed the particle sea in a rather different context: as the state of existence 'before' the Big Bang. From the heat death of our Universe we can therefore expect another universe to spring.

However, our Universe may not expand forever. The galaxies may continue to recede from each other but, under the influence of the Universe's own gravity, begin to slow down and then start to come towards each other again – in exactly the same way, and for exactly the same reason, that a ball thrown into the air will slow and then return to the ground. This view is known as the Oscillating Universe Theory. It implies that the following sequence occurs:

- a Big Bang
- expansion
- contraction, when gravity eventually overcomes expansion
- a 'Big Crunch' as all the matter in the Universe converges to a single point
- another Big Bang

The crucial question here is this: how much matter does the Universe contain? If there is less than a certain amount, the

LEFT *An artist's impression of our Galaxy, a typical spiral. There are billions of galaxies much like it in the Universe.*

gravitational force will be insufficient to stop a never-ending expansion. If there is more than a certain amount, gravity will in due course draw the galaxies back together.

Estimating the mass of all the matter in the Universe is not an easy task but, in the past few years, many scientists have devoted a great deal of time to it. Until quite recently it seemed quite certain (from estimates of stars' masses, interstellar dust, etc.) that the Universe would continue to expand until it met its heat death. Then cosmologists began to revise their estimates of the percentage of the Universe's mass made up by 'dark matter' which could not be directly observed – including black holes and pulsars. Almost overnight, the generally accepted opinion became that the Universe does indeed oscillate.

This notion has various interesting consequences. First, it implies that our Universe is only one of a never-ending succession. This is not too frightening an idea: we have already come across it in terms of the ever-expanding Universe.

Second, it demands that we think about the direction of time. We are accustomed to the 'arrow of time' pointing in a single direction: from past through present to future. In a shrinking Universe the opposite would presumably would be the case: the 'arrow' would go from future through present to past. This is not to say that history would repeat itself backwards: you would not die and then become old, middle-aged and youthful before finally being absorbed into your mother's womb. Much more alarming is the idea that you could not tell the difference between an expanding and a contracting Universe, and might indeed live in a Universe in which the arrow of time points in the opposite direction from the one you think it does! Such a scenario implies that you 'forget' more and more of the future and 'predict' the past much better. (The tenses are, of course, used loosely.)

A third possibility born from the idea of the oscillating Universe is that there is only a certain 'chunk' of time in which the whole succession of universes can exist. (It is much easier to imagine a single 'chunk' of space in which all possible universes can exist; but we must remember that time is a dimension just like the other three.) This implies that a vast number of universes coexist with our own, each having equal reality. This calls us back to the ideas of alternate universes which we were discussing earlier.

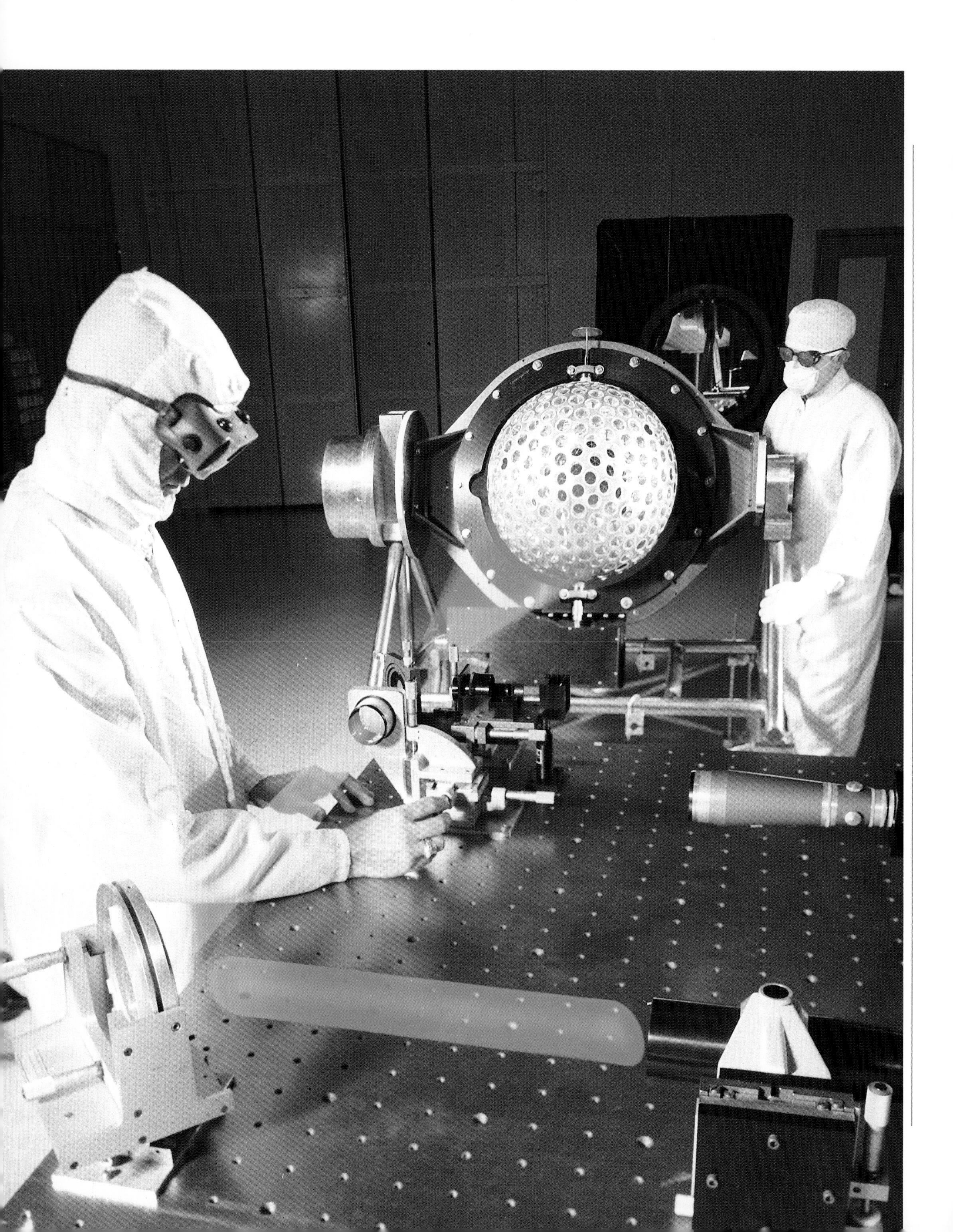

WHAT IS LIGHT 'MADE OF'?

Light is the form of electromagnetic radiation which our eyes are able to see. Other forms of electromagnetic radiation include radio, microwaves and X-rays. The various types of radiation have different wavelengths; if we consider them in terms of increasing or decreasing wavelength, the types of radiation blend gradually from one to the next. The whole range of radiation is called the electromagnetic spectrum.

We tend to think of light as being in some way 'different' from the other forms of radiation, but this is misleading. At a very humble level, the bee can see in the ultraviolet, which is invisible to us: we would be tempted to say that the ultraviolet was not light, because we cannot see it, but as far as the bee is concerned it most certainly is! Similarly, we could postulate some alien being capable of seeing nothing at all in the range of wavelengths we call 'light' but able to see a whole range of colours in the ultraviolet or infrared. The same argument could be extended to any other region of the electromagnetic spectrum (although a creature able to see using, say, X-rays is vanishingly improbable in terms of basic biology). When we use a term like 'light-speed' we are, therefore, really using it as a kind of shorthand to refer to the velocity of any form of electromagnetic radiation. Here we shall use the word 'light' similarly.

The idea that a ray of light is a stream of particles is an old one, dating back at least to ancient Greece. In more recent times it was important for over a century for one reason: it was supported by Sir Isaac Newton. There was already some evidence that light was a form of wave motion, but a wave motion in *what*? (Trying to answer that question would, during the nineteenth century, tie up physics in knots.) Besides, in other ways it was not at all like a wave motion. For example, to compare it with sound, if you play two musical notes that are very close to each other you hear 'beats' – rhythmic loudenings and softenings. This occurs because the waves are not quite in step: sometimes they reinforce each other and at other times they cancel each other out – in the same way that $1 + (-1) = 0$. Light did not seem to do this and so, according to Newton and many others, must be made of particles (because two light particles could never add up to give zero, or darkness).

In 1803 Thomas Young demonstrated a phenomenon called interference. If you have two thin slits, parallel to each other, and shine a very narrow beam of light on them, a barred pattern of alternating dark and bright stripes is projected onto a screen behind. Clearly the two rays of light shining through the slits are having the same sort of effect on each other as the two musical notes described above; this is called interference. Light seemed, therefore, definitely to be a wave motion: and so, it seemed, the debate was over. (Ironically, another interference effect of light is now generally known as Newton's Rings.)

Then, in the early part of this century, the quantum theory appeared; its details need not concern us except for the fact that it implied that all forms of matter and energy came in tiny discrete packages called quanta (by 'discrete' is meant that you cannot have half a quantum). The idea that energy had a quantum nature married neatly enough with

LEFT *Physicists at NASA's Goddard Space Flight Center preparing a satellite for launch. Using laser pulses, the satellite will monitor various tracking stations very accurately in order to give a detailed picture of movements in the Earth's crust and thereby, it is hoped, to help earthquake prediction.*

the view that light was a wave motion, but at the same time it set people thinking in terms of particles once more.

The size of a quantum of light is inversely proportional to the wavelength of the light: the longer the wavelength, the smaller the quantum and the lower the energy of the light.

In the early 1920s the UK physicist Arthur Compton was performing experiments to see what happened when matter was bombarded with X-rays. The beam of X-rays was of course scattered by the collision, but Compton discovered that the wavelengths of some of the scattered X-rays had increased – they had lost energy. How could this be?

Compton proposed that the quantum of light could act as if it were a particle, and he christened this the photon. Photons were colliding with the electrons of the atoms of the matter; the recoil of each electron 'stole' a little energy from the relevant photon – just as the white ball in pool loses energy when it bounces off another ball, and therefore slows down. But photons are extremely odd particles indeed. For one thing, a photon at rest (if you could get one to rest!) has a mass of zero. At light-speed, however, their very velocity gives them the ability to exert pressure. Not very much pressure, to be sure – otherwise all of us would be flattened when we stepped into the sun on a sunny day – but pressure nevertheless.

In passing, we can note that one proposal for spaceflight is to use the pressure of photons from the Sun. The idea is to spread huge sails in space. These would be pushed away by the radiation from the Sun, taking with them any payload attached to them. These solar sails would be slow to accelerate, but could eventually reach high speeds – at zero fuel cost. If sunlight were augmented by powerful lasers, the fuel cost would be much greater but the range of the craft, and their eventual velocity, would be sufficient to allow the technique to be used for interstellar travel.

We can give descriptions of a photon in terms of the way it behaves, but it is very hard for us to imagine what a photon actually *is*. Here we encounter a mystery that is common to all aspects of any discussion of energy or matter at the quantum level. The mathematics works, the evidence can be gathered from experiments, and so forth, but in the end we find ourselves unable in any coherent way to explain the true nature of reality at the quantum level.

LEFT *Albert Einstein with his wife in 1921.*

Can Anything Travel Faster Than Light?

Einstein's Theories of Relativity are generally assumed to tell us that nothing can travel faster than light. In fact, this is not what they say. Rather, they imply that to accelerate any material object up to and beyond light-speed would take an infinite amount of energy and would therefore be impossible. However, they also imply that there could be objects that have *always* travelled at faster-than-light speeds. For such particles, tachyons, infinite energy would be required to *slow them down* to light-speed. Their natural state would be movement at infinite velocity, so that they would be everywhere in the Universe at once.

Everything around us is, by contrast, a tardyon: it is inhibited by the velocity of light. The velocity of light can therefore be seen as a sort of barrier between the Universe we recognize and whatever unimaginable tachyonic universe there might be. This barrier seems to be impermeable: it is unlikely that we shall ever be able directly to detect tachyons (although there are some interesting ideas floating about concerning the consequences were we able to).

Tachyons, if they exist, have some intriguing properties. Perhaps the most interesting is that their time must run in the opposite direction to our own: their 'future' is the birth of the Universe and their 'past' is its end. If future technological societies ever learn to harness tachyons and modulate them (in the same way that we modulate radio waves) they will be able to communicate information from the future to the past. This breaking-down of the time barrier could be extremely exciting; at the same time, the very idea introduces so many paradoxes (if you communicate with the past you change it; you therefore change the present; therefore, in this new present, you never communicated with the past) that we

ABOVE *The young Albert Einstein.*

must think it unlikely that people will ever be able to use tachyons in this way . . . unless communications were flitting between alternate universes, the existence of which would imply the simultaneous reality of all possibilities.

Conversely, these paradoxes may prove to be matters more of perception than of reality. If we imagine ourselves to be standing outside time, looking down upon the history of the Universe as if it were a relief map, we get a different perspective. The future is not interfering with the past; instead, it is fulfilling its obligations, as it were, by bringing about events that have already been reported. In fact, we can turn the apparent paradox on its head: what would happen if the people of the future decided *not* to communicate with the past, despite the fact that their histories told them that they were going to?

Looking at the time-reversal paradoxes again, we see that what they are really referring to is *information*, rather than particles. An individual tachyon whizzing backwards through time is not going to have any discernible effect on the past: it would be only when streams of them were encoded and transmitted to hypothetical 'tachyon detectors' that the paradoxes would begin. The same relationship exists between particles and information when we come to talk about the velocity of light in general. It is easy enough to send information *at* the speed of light: after all, we do exactly this when we wave at a friend or, to be more subtle, use any form of sign language. For theoretical reasons, however, any information sent at a velocity faster than that of light will cause paradoxes similar to those encountered when considering time-reversal. This aspect of Relativity is difficult fully to understand, but in simplest terms we can say that it is impossible to consider whatever is happening at this moment on α Centauri, about 4.5 lightyears away, as sharing a 'now' with whatever is happening here. The 'now' here instead corresponds to the 'now' on α Centauri in 4.5 years' time, because that is how long it takes light to get from here to there. If information could be transmitted from here to α Centauri in less time than this, it would effectively be travelling back into the past.

RIGHT *A photograph of part of the planet* Neptune *taken in* August 1989 *by* Voyager 2, *showing the linear cloud forms.*

A final point to note is that laboratory experiments have confirmed the results of various thought experiments concerning the way that photons behave when they encounter polarizing screens (which, like polaroid sunglasses, allow only photons 'wiggling' in a certain orientation to pass through). The results would seem to confirm that, at least at this level, information *can* be communicated at faster-than-light velocities – instantaneously, in fact. We may have to review exactly what we mean by the word 'paradox' – and, at the same time, rewrite many of our ideas of logic.

Does a Black Hole Orbit the Sun?

Percival Lowell (1855-1916) was not an orthodox astronomer. He was convinced that the surface of the planet Mars was criss-crossed by irrigational canals, betraying the presence on that world of a sophisticated civilization. He was not the first to have been deceived by flaws of telescopic lenses. His other *idée fixe* was, however, a little more rational. At the time it was known that the Sun's retinue consisted of eight planets. The orbit of the outermost of these, Neptune, showed perturbations ('wobbles'), indicating that there was a ninth planet. Lowell instituted a search for this unknown planet, confidently predicting exactly where it should be. His instinct was much better than his maths: some 13 years after his death the planet Pluto was discovered in 1930 by Clyde Tombaugh, using a mixture of perseverance and chance.

There was a mystery, however. Pluto (since discovered to be a 'double planet') is far too small to create the perturbations in Neptune's orbit that started off the search in the first place. The immediate reaction was that the discovery of Pluto was nothing more than a small-scale diversion: there must be a *tenth* planet out there, and a very large planet at that. The search for such a planet has proved unsuccessful – which is something of a mystery, because a planet of that size should be reasonably easy to find. Assuming, that is, that its orbit is not anomalous: one intriguing notion has been that the tenth planet does not orbit the Sun in the same plane as all the others; instead it goes 'up and over'. (Some quite eminent astronomers support this theory.)

However, the Universe is in many ways a very orderly place: if nine planets orbit a star in roughly the same plane, the tenth can be expected to do likewise. Various scientists have therefore proposed that the reason we cannot find the

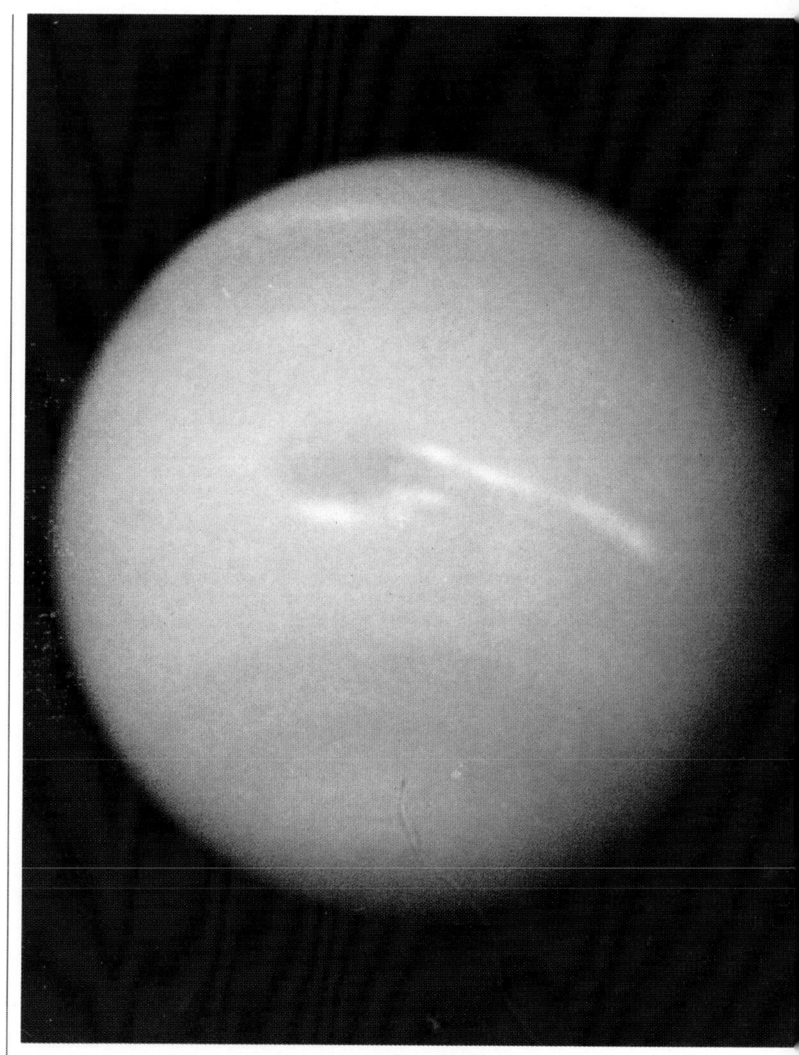

ABOVE *The planet Neptune, as photographed in 1989 by Voyager 2.*

LEFT *The comet Kohoutek photographed in 1973 by Skylab 4.*
ABOVE *Comet Ikeya-Seki photographed by the US Naval Observatory in 1978.*

hypothetical tenth planet is that there is no such planet at all; rather there is, much further out, a black hole. Naturally such an entity would be hard to find, but its gravitational effect would be noticeable.

Our Sun is in a minority among the stars of the Universe in that it seems to be solitary. Most stars are members of double systems, where two stars orbit each other, and systems of three, four, five and even six stars are not uncommon; for example, the star nearest to ours, Proxima Centauri, is the third member of a system whose other two components, orbiting closely together, are jointly called α Centauri. So it would not be surprising if our Sun were in partnership with another star which had degenerated to become a black hole.

There is one difficulty with this. As we have seen, when stars collapse to form black holes an inevitable part of the process is a supernova. The radiation from this explosion would have sterilized all the planets of the Solar System while at the same time creating a huge, expanding shell of gases. However, such a consideration need not obviate the theory. The large stars that become supernovae are very short-lived by comparison with humbler stars like the Sun. The hard radiation from a supernova would be irrelevant to life if the explosion happened before life on Earth had begun to form – and might even have kick-started the process. Likewise, if the explosion occurred early enough in the Solar System's history, the shell of gases could long since have dissipated. A variant of the suggestion is that the Sun has a pulsar as a companion. All of the above considerations apply equally to this situation. One item of evidence in favour of these ideas is the behaviour of comets. It is now generally accepted that, about a lightyear from the Sun, far beyond the orbit of Pluto, there is a belt of at least 10 million comets, collectively known as the Oort Cloud, after the great Dutch astronomer Jan Oort. If the Solar System existed in total isolation, those comets would continue to orbit in perfectly stable fashion. However, the Solar System does not exist in isolation: the gravitational effects of nearby stars are, it is thought, sufficient to deflect some of the comets inwards to swing once – or more often – through the inner Solar System. How much more would a black hole, occasionally passing through or on the outskirts of the Oort Cloud, affect the comets?

If the arrivals of comets in the inner Solar System were totally haphazard we would have little reason to believe such ideas. However, the geological record of past ice ages shows that there is an uncanny regularity in their frequency: they occur approximately every 150 million years. One theory of the origin of ice ages is that they occur when large masses (for example, comets) crash down onto the Earth, throwing up vast amounts of detritus into the stratosphere which block off the heat from the Sun. This periodicity of the ice ages suggests that, once every 150 million years or so, the Earth is particularly vulnerable to cometary bombardment – in other words, that an unusually large number of comets enter the inner Solar System every 150 million years or so. One explanation for this regularity is that, about every 150 million years, a compact companion of the Sun (such as a pulsar or black hole), travelling in a very elliptical orbit, comes close enough to disrupt the comets of the Oort Cloud. What attracts many scientists to this hypothesis is that those perturbations of the orbits of Neptune and Pluto have yet to be explained satisfactorily in any other way.

Why Does the Sun Shine?

Every undergraduate astronomy student can tell you why the Sun shines: a series of nuclear fusion reactions, whose end result is the conversion of hydrogen to helium, happen on a vast scale, and release prodigious amounts of energy in the form of heat, light, X-rays and so on. A part of this reaction should be the release of neutrinos.

Neutrinos are hard to detect: they interact so little with matter that they can probably float through entire galaxies without being affected; they exist but have no mass nor any other physical property, which is like saying that they simultaneously exist and do not exist. Nevertheless, there are various ways by which scientists can detect their presence. The most effective is the use of a very large volume – 450,000 litres (100,000 gallons) – of carbon tetrachloride, better known as a cleaning fluid. In theory at least some neutrinos should react with the chlorine atoms in this mixture as they attempt to pass through it.

The trouble is that they seem not to. If the theory about the way that the Sun shines is correct, the Sun should produce about 180 billion billion billion billion neutrinos each second. Obviously only a small portion of these neutrinos will come in the Earth's direction, but still we ought to be in the path of about 80 billion billion billion neutrinos per second. Clearly, because of the elusive nature of neutrinos we cannot expect to capture all of them. But the results of the best 'neutrino-catching' experiment so far carried out are perhaps a little worse than the expectations of the most depressed pessimist: in 1965 a team of scientists recorded a grand total of seven neutrinos in nine months!

No one is certain why this should be. One possibility is that the Sun is currently going through a 'quiescent' phase: the nuclear reactions at its core are continuing just sufficiently to keep our star shining, but no more than that; in due course the Sun will get back to 'business as usual'. This idea is not very palatable: why should the Sun be anomalous just at the time that we start to search for neutrinos? Alternatively, perhaps neutrinos are much more elusive than we think, so that they evade with ease the traps we set them. A much more likely thesis is that the theory is wrong – either our ideas of the way in which stars function are wildly misplaced or our understanding of the process of nuclear fusion is flawed.

FAR LEFT *This photograph of the comet Kohoutek was taken with a 35mm camera in January 1974 by the lunar and planetary laboratory photographic team at the Catalina Observatory (University of Arizona).* **ABOVE** *Because of scattering in the atmosphere, the Sun's image appears reddened.*

What Causes the Sunspot Cycle?

Whenever you look at a photograph of the Sun you see that the face sports dark blobs. (Technically, the 'face' of the Sun is actually our star's photosphere – the visible level: the Sun has no real surface as such.) These blobs are called sunspots, and they appear dark only because the rest of the Sun is so incandescent: the matter (or plasma, as it should more correctly be called) inside a sunspot, if transported to Earth, would give off greater brilliance than any arc-lamp. The average temperature of the photosphere is about 5400 degrees Celsius; that in a sunspot region is of the order of 3600 degrees Celsius.

Sometimes the face of the Sun is covered in sunspots; other times it is almost naked of them. In the middle of the nineteenth century it was discovered that the highs and lows of sunspot activity recur over a fairly regular cycle of 11 years. In fact the period seems somewhat variable, so that it makes more sense to talk of a 22-year cycle between one sunspot 'high' and, not the next, but the one after that. It seems, therefore, that the prevalence of spots on the Sun's surface follows not a single 11-year cycle but a 22-year 'double-cycle'.

Why should this be so? To date, no one has been able to come up with a satisfactory explanation. An interesting idea proposed many years ago by Fred Hoyle depends upon the proven fact that the Sun has a magnetic field, just like the Earth has. Hoyle suggested that this field's lines of magnetic force manage to twist themselves up every 11 years because

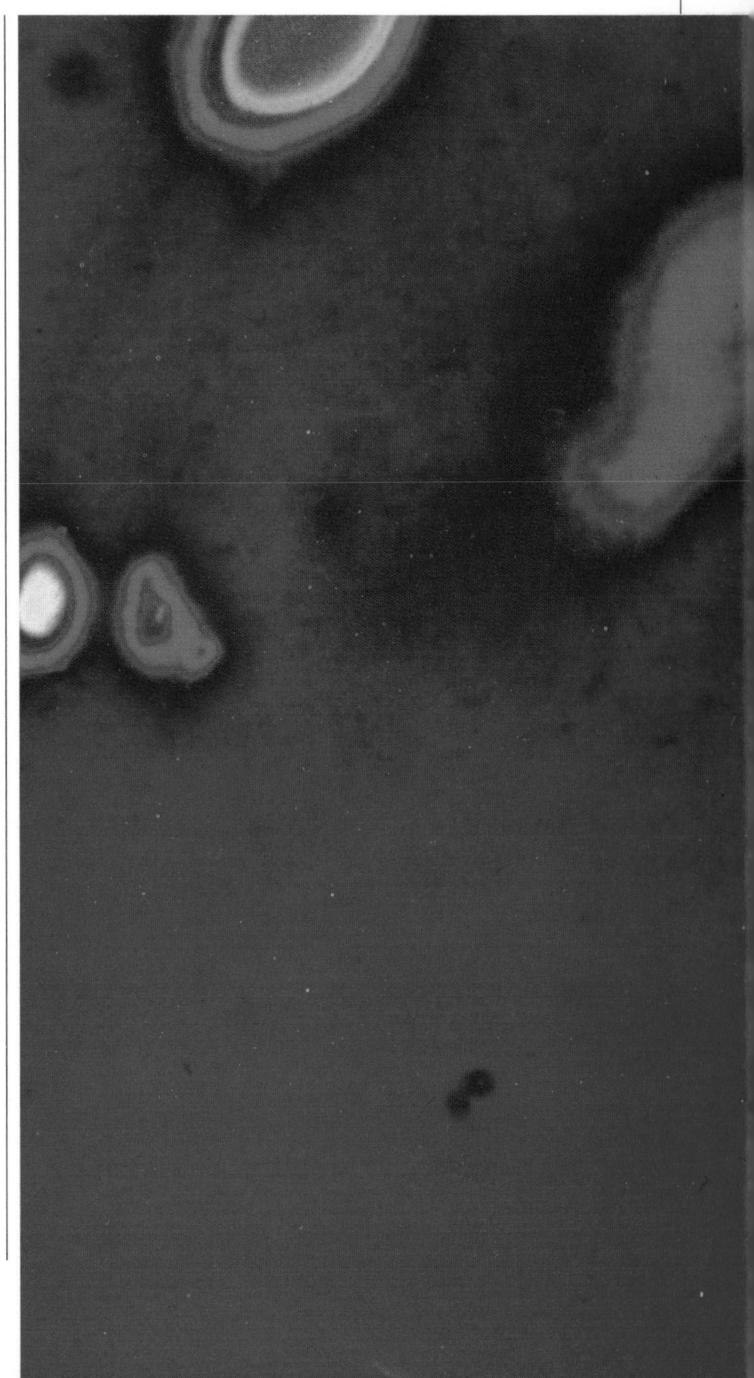

RIGHT *Features of the Sun's corona photographed at X-ray frequencies by Skylab in 1973. The white areas are the brightest and the red the dimmest.*

of the Sun's rapid rotation (once every 34 days or so at the Sun's equator). Hoyle proposed that the magnetic force lines, twisted and contorted, interacted with the Sun's surface in such a way as to produce regions where the nuclear reactions that power the Sun were subdued. Soon thereafter the force lines untwisted – rather like a tangle of elastic bands that had suddenly been released – and matters progressed as normal until, after a further 11 years . . .

The theory is elegant, but unfortunately there is little evidence in its favour. In fact no one has as yet produced a really convincing theory to explain the periodicity of the sunspot cycle, or why it should be periodic. More fundamentally, no one knows why sunspots should occur at all – or perhaps these days we should call them 'starspots', since similar areas of (comparatively) low energy emission have now been detected on the faces of other stars.

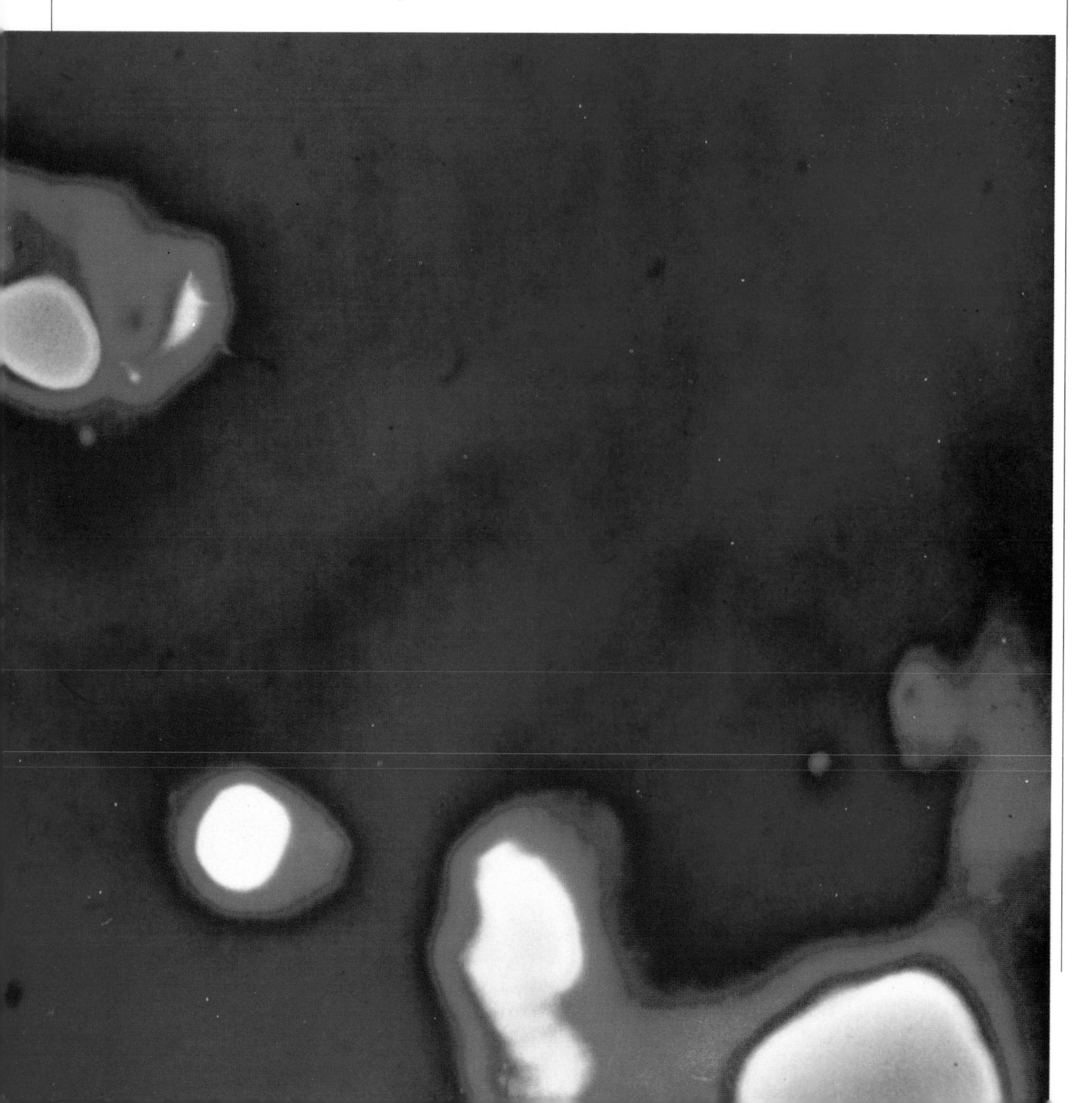

Suggested Further Reading

All of the books in this list are thought-provoking. However, their inclusion here does not mean that the various ideas they propose are valid. So, although these books are certainly worth reading, you would be well advised to treat them all with a fair degree of scepticism.

ALEKSANDER, IGOR, AND BURNETT, PIERS: *Reinventing Man*, London, Kogan Page, 1983.

BAXTER, JOHN, AND ATKINS, THOMAS: *The Fire Came By*, London, Macdonald & Jane's, 1976.

BOYCE, CHRIS: *Extraterrestrial Encounter*, Newton Abbot, David & Charles, 1979.

BRACEWELL, RONALD N: *The Galactic Club*, London, Heinemann, 1978.

CAPRA, FRITJOF: *The Tao of Physics*, London, Wildwood, 1975.

CRICK, FRANCIS: *Life Itself*, London, Macdonald, 1982.

DAVIES, PAUL: *The Edge of Infinity*, London, Dent, 1981.

DAVIES, PAUL: *Other Worlds*, London, Dent, 1980.

DAVIES, PAUL: *Superforce*, London, Heinemann, 1984.

DAWKINS, RICHARD: *The Selfish Gene*, Oxford, OUP, 1976.

DIXON, BERNARD: *Beyond the Magic Bullet*, London, Allen & Unwin, 1978.

DIXON, BERNARD: *What is Science For?*, London, Collins, 1973.

DYSON, FREEMAN: *Disturbing the Universe*, London, Harper & Row, 1979.

EDELSON, EDWARD: *Who Goes There?* New York, Doubleday, 1979.

ELKINGTON, JOHN: *The Gene Factory*, London, Century, 1985.

FRUDE, NEIL: *The Robot Heritage*, London, Century, 1984.

GLEICK, JAMES: *Chaos*, London, Heinemann, 1988.

GOLDSMITH, DONALD, AND OWEN, TOBIAS: *The Search for Life in the Universe*, Menlo, California, Benjamin/Cummings, 1980.

GOOCH, STAN: *Cities of Dreams*, London, Rider, 1989.

GOULD, STEPHEN JAY: *The Mismeasure of Man*, New York, Norton, 1981.

GRANT, JOHN: *A Directory of Discarded Ideas*, Sevenoaks, Ashgrove, 1981.

GRANT, JOHN: *Dreamers*, Bath, Ashgrove, 1984.

GRIBBIN, JOHN: *In Search of Schrödinger's Cat*, London, Wildwood, 1984.

GRIBBIN, JOHN: *Timewarps*, London, Dent, 1979.

HARRINGTON, ALAN: *The Immortalist*, St Albans, Granada, 1973.

HERBERT, NICK: *Quantum Reality*, London, Rider, 1985.

HOYLE, FRED, AND WICKRAMASINGHE, NC: *Evolution from Space*, London, Dent, 1981.

HOYLE, FRED, AND WICKRAMASINGHE, NC: *Lifecloud*, London, Dent, 1978.

HUGHES, DAVID: *The Star of Bethlehem Mystery*, London, Dent, 1979.

JASTROW, ROBERT: *Until the Sun Dies*, New York, Norton, 1977.

JOHN, BRIAN (ED.): *The Winters of the World*, Newton Abbot, David & Charles, 1979.

JONAS, DORIS AND JONAS, DAVID: *Other Senses, Other Worlds*, London, Cassell, 1976.

JONES, ROGER: *Physics as Metaphor*, London, Wildwood, 1983.

KRUPP, EC (ED): *In Search of Ancient Astronomies*, London, Chatto & Windus, 1980.

LAWTON, AT: *A Window in the Sky*, Newton Abbot, David & Charles, 1979.

LEAKEY, RICHARD E, AND LEWIN, ROGER: *Origins*, London, Macdonald & Jane's, 1977.

LUNAN, DUNCAN: *Man and the Planets*, Bath, Ashgrove, 1983.

MORGAN, CHRIS: *Future Man*, Newton Abbot, David & Charles, 1980.

MORGAN, CHRIS, AND LANGFORD, DAVID: *Facts and Fallacies*, Exeter, Webb & Bower, 1981.

NARLIKAR, JAYANT V: *Violent Phenomena in the Universe*, Oxford, OUP, 1982.

NICOLSON, IAIN: *Gravity, Black Holes and the Universe*, Newton Abbot, David & Charles, 1981.

NICOLSON, IAIN: *The Road to the Stars*, Newton Abbot, Westbridge, 1978.

PERCY, WALKER: *Lost in the Cosmos*, London, Arena, 1984.

RENFREW, COLIN: *Before Civilization*, London, Cape, 1973.

ROWAN-ROBINSON, MICHAEL: *Cosmic Landscape*, Oxford, OUP, 1979.

SAGAN, CARL: *Broca's Brain*, London, Hodder & Stoughton, 1979.

SAGAN, CARL: *The Dragons of Eden*, New York, Random, 1977.

SAGAN, CARL, AND DRUYAN, ANN: *Comet*, London, Michael Joseph, 1985.

SULLIVAN, WALTER: *We Are Not Alone*, London, Hodder, 1965.

WARLOW, PETER: *The Reversing Earth*, London, Dent, 1982.

WEINBERG, STEVEN: *The First Three Minutes*, London, Deutsch, 1977.

WHITE, BURTON L: *The First Three Years of Life*, London, WH Allen, 1978.

WHITROW, GJ: *The Nature of Time*, Harmondsworth, Penguin, 1975.

WILSON, COLIN: *Starseekers*, Garden City, Doubleday, 1980.

WILSON, COLIN, AND GRANT, JOHN (EDS.): *The Book of Time*, Newton Abbot, Westbridge, 1980.

WILSON, COLIN, AND GRANT, JOHN: *The Directory of Possibilities*: Exeter, Webb & Bower, 1981.

INDEX

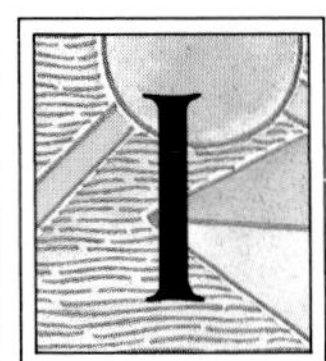

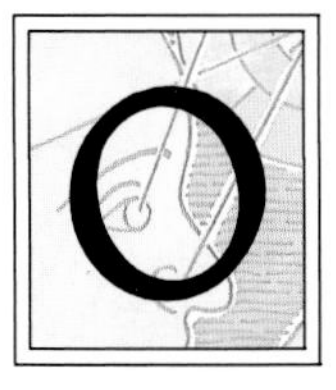